CONTEMPORARY ISSUES IN WIRELESS COMMUNICATIONS

Edited by **Mutamed Khatib**

Contemporary Issues in Wireless Communications
http://dx.doi.org/10.5772/58482
Edited by Mutamed Khatib

Contributors

Adão Silva, Eduardo Castaneda, Atilio Gameiro, Unai Hernandez-Jayo, Sotirios Goudos, Hamed Farhadi, Majid Khormuji, Nima Najari Moghadam, Per Zetterberg, Mikael Skoglund, Vyacheslav Tuzlukov, Abdulsalam Alkholidi, Khalil Altowij

Published by InTech

Janeza Trdine 9, 51000 Rijeka, Croatia

Publishing Process Manager Iva Lipovic

Technical Editor InTech DTP team

Cover InTech Design team

First published December, 2014

Printed in Croatia

Additional hard copies can be obtained from orders@intechopen.com

Contemporary Issues in Wireless Communications, Edited by Mutamed Khatib

p. cm.

ISBN 978-953-51-1732-2

For EU product safety concerns:
IN TECH d.o.o., Prolaz Marije Krucifikse Kozulić 3, 51000 Rijeka, Croatia,
info@intechopen.com or visit our website at intechopen.com.

Meet the editor

Dr. Mutamed Khatib was born in 1973. He received his B.Sc. degree in Telecommunication Engineering from Yarmouk University, Irbid, Jordan in 1996 and M.Sc. in Electrical and Electronic Engineering from Jordan University of Science and Technology, Irbid, Jordan in 2003. He received his PhD Degree in wireless and mobile systems from University Sains Malaysia (USM), Malaysia in 2009. From 1996 to 2005 he worked as Transmission, Outside Broadcasting and Studio Engineer in Palestinian Broadcasting Corporation (PBC). From 2005 to 2009 he worked as an Instructor in the Department of Electrical Engineering, Palestine Technical University (Kadoorie), Tul Karm, Palestine. Since September 2009 Dr. Mutamed Khatib is working as Assistant professor at the same university. Nowadays, he is the dean of college of engineering and technology there.

Dr. Khatib has a number of publications to his credit in various international journals and conference proceedings. He is a member of IEEE, Palestinian Engineers Association and Arab Engineers Association.

Contents

Preface

Wireless communications have seen large growth in the last several years. The huge uptake rate of mobile phone technology, Wireless Local Area Networks (WLAN) and the exponential growth of the Internet have resulted in an increased demand for new methods of obtaining high capacity wireless networks. All of the current communication systems have adopted digital technology. The wireless communication systems are expected to play a more important role in providing portable access to future information services. The demand for new services to support Internet and advanced video applications presents the key technical challenges, i.e., multimedia access requires high-bandwidth and low-latency network connections for many users, mobility requires adaptation to time varying channel conditions and portability imposes severe constraints on receiver size and power consumption. Physical limitations on wireless channels impose huge challenges on reliable communication. Bandwidth limitations, propagation loss, noise and interference make the wireless channel a narrow pipe that does not readily accommodate rapid flow of data.

Thus, researchers aim to design systems that are suitable to operate in such channels, in order to have high performance quality of service. Also, the mobility of the communication systems requires further investigation to reduce the complexity and the power consumption of the receiver.

This book presents new techniques that improve the performance of the communication system used for transmission of digital data over time varying channels such as high frequency mobile channels. It aims to provide highlights of the current research in the field of wireless communication, and to offer a contribution to the recent advances in this field. The subjects discussed in this work are very valuable to any researcher in the communication field, not only to researchers in the wireless related areas. The six chapters cover a wide range of topics in wireless communication starting with evolutionary algorithms, the conventional and adaptive precoders design, ad hoc networks, optical communication, signal processing and interference alignments.

The editor would like to thank all of the authors for their valuable contributions in the area of wireless communications hoping that the book will be of great help to the readers.

Dr. Mutamed Khatib
Dean, College of Engineering and Technology
Palestine Technical University – Kadoorie (PTU)
Tul Karm, Palestine

Chapter 1

Evolutionary Algorithms for Wireless Communications — A Review of the State-of-the art

Sotirios K. Goudos

Additional information is available at the end of the chapter

http://dx.doi.org/10.5772/59147

1. Introduction

Several evolutionary algorithms (EAs) have emerged in the past decade that mimic biological entities behavior and evolution. Darwin's theory of evolution is the major inspiration source for EAs. The foundation of Darwin's theory of evolution is natural selection. The study of evolutionary algorithms began in the 1960s. Several researchers independently developed three mainstream evolutionary algorithms, namely, genetic algorithms [1, 2], evolutionary programming [3], and evolution strategies [4]. EAs are widely used for the solution of single and multi-objective optimization problems. Swarm Intelligence (SI) algorithms are also a special type of EAs. SI can be defined as the collective behavior of decentralized and self-organized swarms. SI algorithms among others include Particle Swarm Optimization (PSO) [5], Ant Colony Optimization [6], and Artificial Bee Colony (ABC) [7].

PSO is an evolutionary algorithm that mimics the swarm behavior of bird flocking and fish schooling [5]. The most common PSO algorithms include the classical Inertia Weight PSO (IWPSO) and Constriction Factor PSO (CFPSO) [8]. PSO is an easy to implement algorithm with computational efficiency. The PSO algorithm is inherently used only for real-valued problems. An option to expand PSO for discrete valued problems also exists. Among others PSO algorithms include, the barebones (BB) and the exploiting barebones (BBExp). BBPSO has been successfully applied to the cell to switch assignment problem [9].

Artificial Bee Colony (ABC) [7] is a recently proposed SI algorithm, which has been applied to several real world engineering problems. The ABC algorithm models and simulates the honey bee behavior in food foraging. In the ABC algorithm, a potential solution to the optimization problem is represented by the position of a food source while the food source corresponds to the quality (objective function fitness) of the associated solution. The ABC algorithm has been successfully applied to several problems in wireless communications [10]. ABC variants that improve the original algorithm have also been proposed [11].

Ant Colony Optimization (ACO) is a population-based metaheuristic introduced by Marco Dorigo [12]. This algorithm was inspired by the behaviour of real ants. The algorithm is based on the fact that ant colonies can find the shortest path between their nest and a food source just by depositing and reacting to pheromones while they are exploring their environment. ACO is suitable for solving combinatorial optimization problems, which are common in wireless communications.

Differential evolution (DE) [13, 14] is a population-based stochastic global optimization algorithm, which has been used in several real world engineering problems. Several DE variants or strategies exist. One of the DE advantages is that very few control parameters have to be adjusted in each algorithm run. However, the control parameters involved in DE are highly dependent on the optimization problem. Moreover, the selection of the appropriate strategy for trial vector generation requires additional computational time using a trial-and-error search procedure. Therefore, it is not always an easy task to fine-tune the control parameters and strategy. Since finding the suitable control parameter values and strategy in such a way is often very time-consuming, there has been an increasing interest among researchers in designing new adaptive and self-adaptive DE variants. Self adaptive DE (SaDE), is a DE algorithm that self-adapts both control parameters and strategy based on learning experiences from previous generations is presented in [15-17]. SaDE has been applied to microwave filter design, [18], and to antenna arrays synthesis [19].

The purpose of this chapter is to briefly describe the above algorithms and present their application to wireless communications optimization problems found in the literature. This chapter also presents results from different cases using PSO, ABC, ACO and DE. These include the cell to switch assignment problem in cellular networks using PSO algorithms, peak to average power ratio (PAPR) reduction of OFDM signals with the partial transmit sequences (PTS) approach using ABC and ACO algorithms [7, 11], and dual-band microwave filter design for wireless communications using SADE.

This chapter is subdivided into four sections. Section 2 presents the different evolutionary algorithms. Section 3 reviews the related work in wireless communications problems from the literature. Section 4 describes the design cases and presents the numerical results. Finally section 5 contains the discussion about the advantages of using a EA-based approach and the conclusions.

2. Methods

A population (or swarm) in PSO, ABC, ACO and DE consists of NP vectors (or particles) $\bar{x}_{G,i}$, i=1, 2, ..., NP, where G is the generation number. The population is initialized randomly from a uniform distribution. Each D-dimensional vector represents a possible solution, which is expressed as:

$$\overline{x}_{G,i} = (x_{G,1i}, x_{G,2i}, \ldots x_{G,ji}, \ldots\ldots, x_{G,Di}) \quad (1)$$

The population is initialized as follows:

$$x_{0,ji} = rand_{j[0,1)}\left(x_{j,U} - x_{j,L}\right) + x_{j,L} \quad j = 1,2,\ldots\ldots,D \quad (2)$$

where $x_{j,L}$ and $x_{j,U}$ are *D*-dimensional vectors of the lower and upper bounds respectively and $rand_{j[0,1)}$ is a uniformly distributed random number within [0,1). The stopping criterion for PSO, ABC and DE is usually the generation number or the number of objective-function evaluations.

2.1. Particle Swarm Optimization (PSO)

In PSO, the particles move in the search space, where each particle position is updated by two optimum values. The first one is the best solution (fitness) that has been achieved so far. This value is called *pbest*. The other one is the global best value obtained so far by any particle in the swarm. This best value is called *gbest*. After finding the *pbest* and *gbest*, the velocity update rule is an important factor in a PSO algorithm. The most commonly used algorithm defines that the velocity of each particle for every problem dimension is updated with the following equation:

$$u_{G+1,ni} = wu_{G,ni} + c_1 rand_{1(0,1)}(pbest_{G+1,ni} - x_{G,ni}) + c_2 rand_{2(0,1)}(gbest_{G+1,ni} - x_{G,ni}) \quad (3)$$

where $u_{G+1,ni}$ is the ith particle velocity in the nth dimension, *G+1* denotes the current iteration and G the previous, $x_{G,ni}$ is the particle position in the nth dimension, $rand_{1(0,1)}$, $rand_{2(0,1)}$ are uniformly distributed random numbers in (0,1), w is a parameter known as the inertia weight, and c_1 and c_2 are the learning factors.

The parameter w (inertia weight) is a constant between 0 and 1. This parameter represents the particle's fly without any external influence. The higher the value of w, or the closer it is to one, the more the particle stays unaffected from pbest and gbest. The inertia weight controls the impact of the previous velocity: a large inertia weight favors exploration, while a small inertia weight favors exploitation. The parameter c_1 represents the influence of the particle memory on its best position, while the parameter c_2 represents the influence of the swarm best position. Therefore, in the Inertia Weight PSO (IWPSO) algorithm the parameters to be determined are: the swarm size (or population size), usually 100 or less, the cognitive learning factor c_1 and the social learning factor c_2 (usually both are set to equal to 2.0), the inertia weight w, and the maximum number of iterations. It is common practice to linearly decrease the inertia weight starting from 0.9 or 0.95 to 0.4.

Clerc [8] suggested the use of a different velocity update rule, which introduced a parameter K called constriction factor. The role of the constriction factor is to ensure convergence when all the particles have stopped their movement. The velocity update rule is then given by:

$$u_{G+1,ni} = K\left[u_{G,ni} + c_1 rand_{1(0,1)}(pbest_{G+1,ni} - x_{G,ni}) + c_2 rand_{2(0,1)}(gbest_{G+1,ni} - x_{G,ni})\right] \tag{4}$$

$$K = \frac{2}{\left|2 - \varphi - \sqrt{\varphi^2 - 4\varphi}\right|} \tag{5}$$

where $\varphi = c_1 + c_2$ and $\varphi > 4$.This PSO algorithm variant is known as Constriction Factor PSO (CFPSO).

2.2. Barebones PSO

Kennedy [20] proposed a new PSO approach, the BB PSO, where the standard PSO velocity equation is replaced with samples from a normal distribution. In this method, the position update rule for the ith particle in the nth dimension becomes

$$x_{G+1,ni} = N\left(\frac{pbest_{G+1,ni} + gbest_{G+1,ni}}{2}, \left|pbest_{G+1,ni} - gbest_{G+1,ni}\right|\right) \tag{6}$$

$N(\, , \,)$ denotes the normal distribution. The method allows particles with *pbest* significant different from *gbest* to make large step sizes towards it. When *pbest* is close to *gbest*, step size decreases and limits exploration in favor of exploitation.

In [20], a variation of BB PSO, the BBExp PSO, was also proposed. In this method, approximately half of the time velocity is based on samples from a normal distribution; for the rest of the time, velocity is derived from the particle's personal best position. The position update rule, (6), is modified into

$$x_{G+1,ni} = \begin{cases} N\left(\frac{pbest_{G+1,ni} + gbest_{G+1,ni}}{2}, \left|pbest_{G+1,ni} - gbest_{G+1,ni}\right|\right), & U(0,1) > 0.5 \\ pbest_{G+1,ni}, & \text{otherwise} \end{cases} \tag{7}$$

where $U(\, , \,)$ denotes the uniform distribution. In BBExp PSO, position updates equal $pbest_n$ for half of the time resulting in the improved exploitation of $pbest_n$ compared to the BB PSO. One may notice

that the barebones PSO algorithms do not require parameter tuning. More details can be found in [20, 21].

2.3. Artificial bee colony optimization

The ABC algorithm models and simulates the honey bee behavior in food foraging. In ABC algorithm, a potential solution to the optimization problem is represented by the position of a food source while the nectar amount of a food source corresponds to the quality (objective function fitness) of the associated solution. In order to find the best solution the algorithm defines three classes of bees: employed bees, onlooker bees and scout bees. The employed bee searches for the food sources, the onlooker bee makes a decision to choose the food sources by sharing the information of employed bee, and the scout bee is used to determine a new food source if a food source is abandoned by the employed bee and onlooker bee. For each food source exists only one employed bee (i.e. the number of the employed bees is equal to the number of solutions). The employed bees search for new neighbor food source near of their hive. A new position of the $\bar{x}_i = (x_{i,1}, .., x_{i,j}, .., x_{i,D})$ solution, where D is the problem dimension, is generated using

$$u_{i,j} = x_{i,j} + \varphi_{i,j}\left(x_{i,j} - x_{k,j}\right) \tag{8}$$

where $k \in \{1, 2, .., SN\}$, $k \neq i$, $j \in \{1, 2, .., D\}$ are randomly chosen indices, where SN is the number of food sources, and $\varphi_{i,j}$ is a uniformly distributed random number within [-1,1]. ABC uses a greedy selection operator, which for minimization problems is defined by

$$\bar{x}_i' = \begin{cases} \bar{u}_i, & \text{if } f(\bar{u}_i) < f(\bar{x}_i) \\ \bar{x}_i, & \text{otherwise} \end{cases} \tag{9}$$

where $\bar{x}_i'$ is the new position of the food source.

An onlooker bee chooses a food source depending on the probability value associated with that food source, p_i, given by

$$p_i = \frac{fit_i}{\sum_{m=1}^{SN} fit_m} \tag{10}$$

where fit_i is the fitness value of the ith solution which is proportional to the nectar amount of the food source in the ith position. When a food source (solution) cannot be improved anymore then the scout bee helps the colony to randomly generate create new solutions

$$x_{i,j} = rand_{j(0,1)}\left(x_{j,U} - x_{j,L}\right) + x_{j,L} \quad j = 1, 2, ..., D \tag{11}$$

where $x_{j,L}$ and $x_{j,U}$ are the lower and upper bounds of the jth dimension respectively and $rand_{j(0,1)}$ is a uniformly distributed random number within (0,1).

2.4. Ant colony optimization

Ant colony optimization (ACO) [6, 12, 22] is a meta-heuristic inspired by the ants' foraging behavior. At the core of this behavior is the indirect communication between the ants by means of chemical pheromone trails, which enables them to find short paths between their nest and food sources. Ants can sense pheromone. When they decide path to follow a path, they tend to choose the ones with strong pheromone intensities way back to the nest or to the food source. Therefore, shorter paths would accumulate more pheromone than longer ones. This feature of real ant colonies is exploited in ACO algorithms in order to solve combinatorial optimization problems considered to be NP-Hard.

2.5. Differential Evolution (DE)

The initial population evolves in each generation with the use of three operators: mutation, crossover and selection. Depending on the form of these operators several DE variants or strategies exist in the literature [14, 23]. The choice of the best DE strategy depends on problem type [24]. In SaDE the following four strategies are used for trial vector generation. These include *DE/rand/1bin, DE/rand-to-best/2/bin, DE/rand/2/bin,* and *DE/current-to-rand/1* [25]. In these strategies a mutant vector $\bar{v}_{G+1,i}$ for each target vector $\bar{x}_{G,i}$ is computed by:

$$\begin{aligned}
&DE/rand/1/bin\\
&\bar{v}_{G+1,i} = \bar{x}_{G,r_1} + F(\bar{x}_{G,r_2} - \bar{x}_{G,r_3}),\quad r_1 \neq r_2 \neq r_3\\
&DE/rand-to-best/2/bin\\
&\bar{v}_{G+1,i} = \bar{x}_{G,i} + F(\bar{x}_{G,best} - \bar{x}_{G,i}) + F(\bar{x}_{G,r_1} - \bar{x}_{G,r_2}) + F(\bar{x}_{G,r_3} - \bar{x}_{G,r_4}),\quad r_1 \neq r_2 \neq r_3 \neq r_4\\
&DE/rand/2/bin\\
&\bar{v}_{G+1,i} = \bar{x}_{G,r_1} + F(\bar{x}_{G,r_2} - \bar{x}_{G,r_3}) + F(\bar{x}_{G,r_4} - \bar{x}_{G,r_5}),\quad r_1 \neq r_2 \neq r_3 \neq r_4 \neq r_5\\
&DE/current-to-rand/1/bin\\
&\bar{v}_{G+1,i} = \bar{x}_{G,i} + K(\bar{x}_{G,r_1} - \bar{x}_{G,i}) + F(\bar{x}_{G,r_2} - \bar{x}_{G,r_3}),\quad r_1 \neq r_2 \neq r_3
\end{aligned} \tag{12}$$

where r_1, r_2, r_3, r_4, r_5 are randomly chosen indices from the population, which are different from index i, F is a mutation control parameter, K a coefficient responsible for the level of recombination that occurs between $\bar{x}_{G,i}$ and $\bar{x}_{G,r_1}$. After mutation, the crossover operator is applied to generate a trial vector $\bar{u}_{G+1,i}=(u_{G+1,1i}, u_{G+1,2i}, ...u_{G+1,ji}, ..., u_{G+1,Di})$ whose coordinates are given by:

$$u_{G+1,ji} = \begin{cases} v_{G+1,ji}, & \text{if } rand_{j[0,1)} \leq CR \text{ or } j=rn(i) \\ x_{G+1,ji}, & \text{if } rand_{j[0,1)} > CR \text{ and } j \neq rn(i) \end{cases} \tag{13}$$

where $j=1, 2, \ldots\ldots, D$, $rand_{j[0,1)}$ is a number from a uniform random distribution from the interval [0,1), rn(i) a randomly chosen index from (1, 2, ..., *D*), and *CR* the crossover constant from the interval [0,1]. DE uses a greedy selection operator, which for minimization problems is defined by:

$$\overline{x}_{G+1,i} = \begin{cases} \overline{u}_{G+1,i}, & \text{if } f(\overline{u}_{G+1,i}) < f(\overline{x}_{G,i}) \\ \overline{x}_{G,i}, & \text{otherwise} \end{cases} \tag{14}$$

where $f(\bar{u}_{G+1,i})$, $f(\bar{x}_{G,i})$ are the fitness values of the trial and the old vector respectively. Therefore, the newly found trial vector $\bar{u}_{G+1,i}$ replaces the old vector $\bar{x}_{G,i}$ only when it produces a lower objective-function value than the old one. Otherwise, the old vector remains in the next generation. The stopping criterion for the DE is usually the generation number or the number of objective-function evaluations.

2.6. Self-Adaptive DE (SADE)

In the SaDE algorithm both the trial vector generation strategies and the control parameters are self-adapted according to previous learning experiences. SaDE maintains a strategy candidate pool, consisting of the four strategies given in (2). Each strategy is assigned a certain probability. The sum of all probabilities is equal to one. These probabilities are initialized with a value of 0.25 and gradually adapted during evolution. The probability of applying the *m-th* strategy is p_m, $m=1, 2, \ldots\ldots, M$, where *M* is the total number of strategies. At generation *G*, the number of successful trial vectors generated by the *m-th* strategy is denoted as $ns_{m,G}$, while the number of trial vectors that fail to replace the old vectors in the next generation is $nf_{m,G}$. An additional parameter called the learning period (LP) is introduced in [17]. This corresponds to the number of the previous generations that store the success and fail statistics. After LP generations, the probabilities of selecting different strategies are updated according to:

$$p_{m,G} = \frac{S_{m,G}}{\sum_{m=1}^{M} S_{m,G}}$$

where

$$S_{m,G} = \frac{\sum_{g=G-LP}^{G-1} ns_{m,g}}{\sum_{g=G-LP}^{G-1} ns_{m,g} + \sum_{g=G-LP}^{G-1} nf_{m,g}} + \varepsilon \tag{15}$$

where $S_{m,G}$ is the success rate of the trial vectors generated by the *m-th* strategy within the previous LP generation and ε is a constant set equal to 0.01 to avoid possible null success rates.

Therefore, according to (5) strategies with high success rates have higher probability to be applied at the current generation.

The control parameters are self-adapted in the following way. The mutation control parameter F is approximated by a normal distribution with mean value 0.5 and standard deviation 0.3, that is $N(0.5, 0.3)$. The parameter K is a random number in the interval [0, 1] generated by a uniform distribution. The crossover rate control parameter CR used by the *m-th* strategy is also approximated by a normal distribution with mean value CR_m and standard deviation 0.1, that is $N(CR_m, 0.1)$. The initial value of CR_m is 0.5 for all strategies. The values of crossover rates that have successfully generated trial vectors in the previous LP generations are stored in a crossover rate memory for each strategy CR_m^{memory} that is an array of size LP. At each generation, the median value stored in memory for the *m-th* strategy CR_m^{median} is calculated and the CR values generated are given by a normal distribution with mean value CR_m^{median} and standard deviation 0.1. That way the crossover values are evolved at each generation to follow the successful values found. The authors in [17] suggest a value between 20 and 60 for the parameter LP. The sensitivity analysis performed in [17] for the LP parameter showed it had no significant impact on SaDE performance. More details about the SaDE algorithm can be found in [17].

3. Related work

This section presents a brief literature review of applications of evolutionary algorithms and their variants to wireless communications problems.

Genetic algorithms (GAs) are among the widely used optimization techniques for addressing design problems in wireless communications. In [26] a Smith prediction filter is proposed for power control design of direct-sequence code-division multiple-access cellular mobile radio systems. A fixed-order robust $H\infty$ loop filter is developed using a genetic algorithm to minimize the worst-case variance of the received SINR from the minimax perspective. The authors in [27] present an antenna selection method for multiple-input multiple-output wireless systems based on a GA that seeks the best subset of antenna elements. The problem of receive antenna selection and symbol detection for multiple-input, multiple-output (MIMO) systems is solved in [28] by applying a genetic algorithm (GA) variant. The paper in [29] addresses the problem of joint transmit/receive antenna selection for MIMO systems using a real-valued genetic algorithm (RVGA). The optimization objective is to improve the channel capacity of multiple-input/multiple-output (MIMO) systems. The study in [30] presents a pattern discovery algorithm for multi-streams mining in wireless sensor networks. This algorithm adapts genetic operators with Elitism Strategy.The paper in [31] studies joint multiuser linear precoding design in the forward link of fixed multibeam satellite systems. The authors use a generic optimization framework for linear precoding design to handle any objective functions of data rate with general linear and nonlinear power constraints. In [32] an energy-efficient genetic algorithm mechanism is presented to resolve quality of service (QoS) multicast routing problem: The proposed genetic algorithm depends on bounded end-to-end delay and minimum energy cost of the multicast tree.

SI algorithms are among the most commonly used algorithms for solving problems in wireless communications. PSO and several PSO variants have been used in the literature to solve different problems. In [33] optimal power scheduling for distributed detection in a Gaussian sensor network is addressed for both independent and correlated observations. A PSO based technique is developed to find the optimal power allocation for arbitrary correlations. The authors in [34] apply PSO to solve the constrained nonlinear optimisation problem for the minimum bit error rate (MBER) multiuser transmitter (MUT). The proposed PSO aided symbol-specific MBER-MUT and average MBER-MUT schemes provide improved performance in comparison to the conventional minimum mean-square error MUT scheme. Several issues in Wireless Sensor Networks (WSNs) can be formulated as multidimensional optimization problems, and addressed using bio-inspired algorithms. In [35] the authors present a brief survey of how PSO is used to address these issues. The authors in [36] propose a new approach to estimate the location of a sensor in a wireless sensor network based on a new PSO algorithm with a log-barrier approach. The paper in [37] presents a predistorter based on a cluster-based implementation particle swarm optimization technique with embedded model-size estimation capability and validates the proposed technique on a Doherty power amplifier prototype.

The ABC algorithm and its variants have been successfully applied to several optimization problems in wireless communications. Among others these include issues in WSN [38-41], WiMax network planning [42] and channel assignment [43].

The ACO algorithm has also been applied to several combinatorial problems in wireless systems which include problems in mobile ad hoc networks [44-46], problems in WSN [47-51], cognitive radio [52], resource allocation in multiuser OFDM systems [53] and MIMO problems [54, 55].

DE variants have also been applied to variety of optimization problems like multi-user detection in multi-carrier CDMA [56], WSNs issues [57, 58], urban area path loss prediction [59], spectrum sharing [60], optimization of interleave-division multiple-access communication systems [61]. Other papers use a number of different optimization algorithms and compare results. For example in [62] spectrum allocation methods for cognitive radio based on GA, quantum genetic algorithm (QGA), and PSO, are proposed.

4. Results and discussions

In this section we present numerical results from different optimization problems in wireless communications using different algorithms.

4.1. Cell to switch assignment in cellular networks using barebones PSO

Cell assignment is an important issue in the area of resource management in cellular networks. The problem is an NP-hard one and requires efficient search techniques for its solution in real-time. We briefly present an example case of solving this problem using the barebones PSO [9]. The effective assignment of cells to switches in order to minimize the cost of network deployment is a challenging issue in cellular networking.

The cell-to-switch assignment (CSA) problem consists of optimally assigning cells to network switches while respecting certain constraints such as the call volume of each cell and the switches capacity [63]. The objective of the optimization is the reduction of implementation and operational costs. Usually, the cost function considers the cost of linking cells to switches (cabling cost) and the cost of handoff between different cells (handoff cost). The problem is an NP-hard one with exponential complexity and cannot be solved analytically in real size networks. Other evolutionary techniques like tabu search [64], ACO [65] and GAs [66] have been used in the literature for solving this problem. The problem formulation is given below.

We consider n unique and distinct cells in a given service area and m switches with known location and traffic parameters. The objective is the optimum assignment of cells to switches in order to minimize the total cost that comprises the handoff and cabling costs.

In single-homing CSA, each cell belongs to one cluster and it is assigned to one switch at a time. In this case, the objective function to be minimized is [63]:

$$\sum_{i=1}^{n}\sum_{j=1}^{m} c_{ik}x_{ik} + \sum_{i=1}^{n}\sum_{\substack{j=1 \\ j\neq i}}^{n} h_{ij}\left(1-y_{ij}\right), \quad k=1,\ldots,m \tag{16}$$

where c_{ik} is the cabling cost per time unit between cell i and switch k, x_{ik} is a parameter that takes the value one when cell i is assigned to switch k (otherwise, x_{ik}=0) and h_{ij} is the cost per time unit for the handoffs that occur between cells i and j. The first term in (16) gives the total cabling cost and the second one is the total handoff cost per time unit among cells. Therefore, y_{ij} is defined as

$$y_{ij} = \sum_{k=1}^{m} x_{ik}x_{jk}, \quad i,j=1,\ldots,n \tag{17}$$

where y_{ij} is one when cells i and j are connected to the same switch, otherwise it is zero. The product $x_{ik}x_{jk}$ in (17) defines the variable

$$z_{ijk}=x_{ik}x_{jk} \quad i,j=1,\ldots,n \text{ and } k=1,\ldots,m \tag{18}$$

that is zero unless cells i and j are connected to switch k. In this case, it is one.

Cell assignment is subject to further constraints. The call handling capacity of each switch should not be violated at any time, i.e.

$$\sum_{k=1}^{n} \lambda_i x_{ik} \leq M_k, \quad i = 1,\ldots,n \tag{19}$$

where λ_i is the number of calls that cell i handles per unit time and M_k is the call handling capacity of switch k. Also, each cell is assigned only to one switch, i.e.

$$\sum_{k=1}^{m} x_{ik} = 1, \quad i = 1,\ldots,n \tag{20}$$

The optimization problem defined by (16) and subject to (17)-(20), can be converted [63] to an integer programming one by replacing (18) with the

$$\begin{aligned} &0 \leq z_{ijk} \leq x_{jk}, x_{ik} \\ &z_{ijk} \geq x_{ik} + x_{jk} - 1, \end{aligned} \quad i,j = 1,\ldots,n \text{ and } k = 1,\ldots,m \tag{21}$$

We compare BB and BBExp PSO with ACO [65] and binary PSO (BPSO) [67]. We run 100 independent trials for each algorithm. The statistical results are presented and compared In the proposed barebones PSO variants, the only parameter we set was the swarm size. In the examples presented here, this was set to 5 particles. The ACO parameters were the same as in [65]. For the BPSO, we set the learning factors c_1 and c_2 equal to two. Systems with varied number of cells and switches that range from 15 to 200 and from 2 to 7, respectively, were considered.

The percentage of successfully obtained solutions as a function of the number of cells and switches indicates the solution accuracy of the algorithms. Figure 1 shows the results of the application of BB PSO, BBExp PSO, BPSO and ACO in a single-homing system for swarm size equal to 5 particles. In the first case, BB PSO outperforms the other methods in systems with small complexity but as the complexity increases (n/m=150/6 and 200/7) BBExp gives the best results. As it was expected, solution accuracy decreases with system complexity. In general, BB PSO outperforms the other methods for small n and m. However, its performance degrades with system complexity; in this case, BBExp gives better results. In any case, at least one of BB and BBExp PSO is better than BPSO and ACO. We have also evaluated the different algorithms in terms of the computational time required for the derivation of the previous results. Figure 2 shows small differences in results between the four algorithms. In all cases, barebones PSO algorithms are slightly faster. BBExp outperforms BB as system complexity increases. The computational cost of the methods grows exponentially with n/m and increases with the swarm size. More details about this problem can be found in [9].

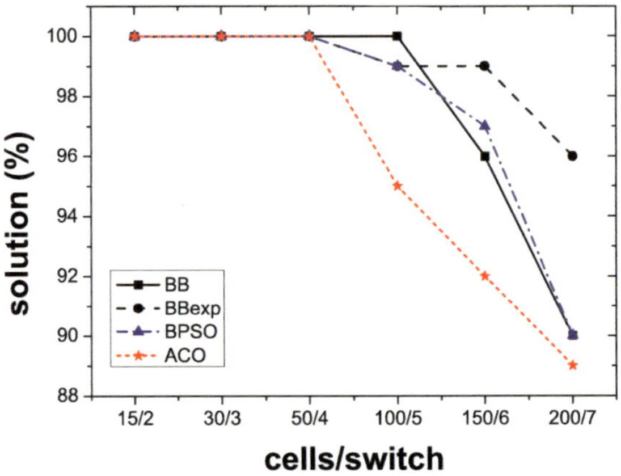

Figure 1. Successful solutions vs cells/switch

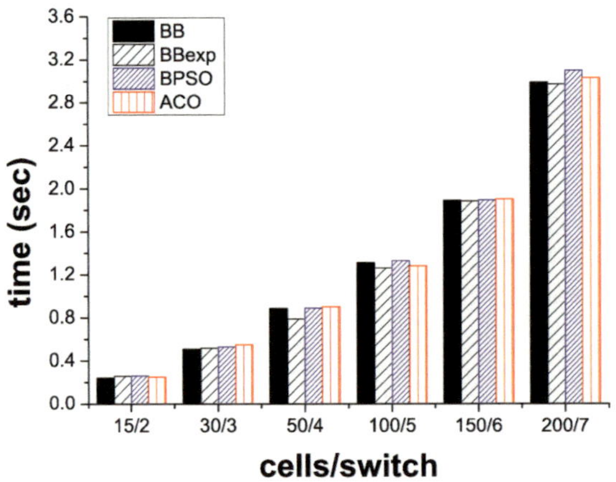

Figure 2. Computational time vs cells/switch chart

4.2. A PTS technique based on ACO and ABC for PAPR reduction of OFDM signals

A major drawback of OFDM signals is the high value of peak to average power ratio (PAPR). Partial transmit sequences (PTS) [68], is a popular PAPR reduction method with good PAPR reduction performance. However, PTS requires an exhaustive search in order to find the optimal phase factors. Thus, the search complexity is high. Several methods have been published in the literature for PAPR reduction using PTS with low search complexity [10, 69, 70]. The problem description is given below.

In an OFDM system, the high-rate data steam is split into N low-rate data streams that are simultaneously transmitted using N subcarriers. The discrete-time signal of such a system is given by

$$s_k = \frac{1}{\sqrt{N}} \sum_{n=0}^{N-1} S_n e^{\frac{j2\pi nk}{LN}} \quad k = 0,1,\ldots.,LN-1 \tag{22}$$

where L is the oversampling factor, $\boldsymbol{S} = [S_0, S_1, \ldots, S_{N-1}]^T$ is the input signal block. Each symbol is modulated by either phase-shift keying (PSK) or quadrature amplitude modulation (QAM).

The PAPR of the signal in (22) is defined as the ratio of the maximum to average power and is expressed in dB as

$$PAPR(s) = 10\log_{10} \frac{\max\limits_{0 \le k \le LN-1} |s_k|^2}{E\left[|s_k|^2\right]} \tag{23}$$

where $E[.]$ is the expected value operation.

In the PTS approach the $\boldsymbol{S}$ input data OFDM block is partitioned into M disjointed subblocks represented by the vector $\boldsymbol{S}_m$ $m=1, 2, \ldots, M-1$ and oversampled by inserting $(L-1)N$ zeros. Then the PTS process is expressed as

$$\boldsymbol{S} = \sum_{m=1}^{M} \boldsymbol{S}_m \tag{24}$$

Next, the subblocks are converted to time domain using LN point inverse fast fourier transform (IFFT). The representation of the OFDM block in time domain is expressed by

$$\boldsymbol{s} = IFFT\left\{\sum_{m=1}^{M} \boldsymbol{S}_m\right\} = \sum_{m=1}^{M} IFFT\{\boldsymbol{S}_m\} = \sum_{m=1}^{M} \boldsymbol{s}_m \tag{25}$$

The PTS objective is to produce a weighted combination of the M subblocks using $\boldsymbol{b} = [b_1, b_2, \ldots, b_M]^T$ complex phase factors to minimize PAPR. The transmitted signal in time domain after this combination is given by

$$\boldsymbol{s}'(\boldsymbol{b}) = \sum_{m=1}^{M} b_m \boldsymbol{s}_m \tag{26}$$

In order to reduce the search complexity the phase factor possible values are limited to a finite set. The set of allowable phase factors is

$$A = \left\{ e^{\frac{j2\pi n}{W}} \middle| n = 0,1,\ldots,W-1 \right\} \tag{27}$$

where W is the number of allowed phase factors. Therefore in case of M subblocks and W phase factors the total number of possible combinations is W^M. In order to reduce the search complexity we usually set fixed one phase factor.

The optimization goal of the PTS scheme is to find the optimum phase combination for minimum PAPR. Thus, the objective function can be expressed as

Minimize

$$F(\boldsymbol{b}) = 10\log_{10} \frac{\max\limits_{0 \le k \le LN-1} \left| s'(\boldsymbol{b}) \right|^2}{E\left[\left| s'(\boldsymbol{b}) \right|^2 \right]} \tag{28}$$

subject to

$$\boldsymbol{b} \in \left\{ e^{j\phi_m} \right\}^M \text{ where } \phi_m \in \left\{ \frac{2\pi l}{W} \middle| l = 0,1,\ldots,W-1 \right\} \tag{29}$$

We have evaluated objective function above using evolutionary algorithms and methods found in the literature. We have used two main measurement criteria namely the complementary cumulative distribution function (CCDF) and the computational complexity. In all our simulations, 10E5 random OFDM blocks are generated. The transmitted signal is oversampled by a factor L=4. We consider 16-QAM modulation with N=256 sub-carriers which are divided into M=16 random subblocks. The phase factors for W=2 are selected. We consider the first phase factor to be fixed so the total number of unknown phase factors is M-1.

The control parameters in all simulations are given below. In the PSO algorithm c_1 and c_2 are set equal to 2.05 while the inertia weight is linearly decreased starting from 0.9 to 0.4. For ACO the initial pheromone value τ_0 is set to 1.0e-6, the pheromonone update constant Q is set to 20, the exploration constant q_0 is set to 1, the global pheromone decay rate ρ_g is 0.9, the local pheromone decay rate ρ_l is 0.5, the pheromone sensitivity α is 1, and the visibility sensitivity is β is 5.

Figure 3 presents the comparison between the CCDF by different PTS reduction techniques. For $\Pr(PAPR > PAPR_0) = 10^{-3}$ the PAPR of the original OFDM transmitted signal is 10.84dB. For all evolutionary algorithms, the population size NP is set to 30 and the maximum number of

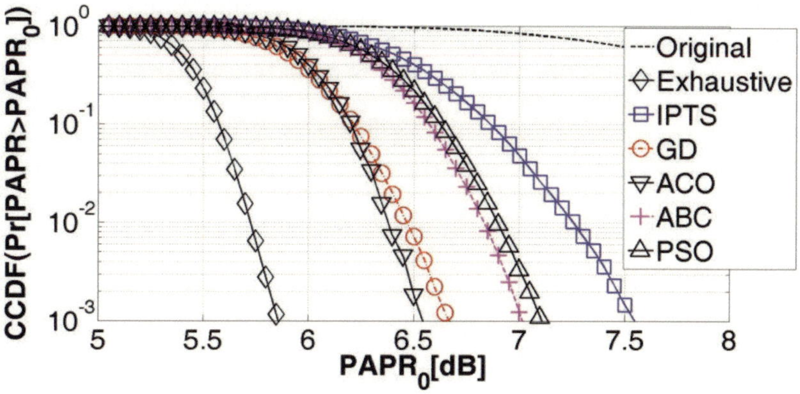

Figure 3. PARP reduction performance comparison of the BBO-PTS algorithms with other PTS schemes for NP=30, G=30.

generations G is set to 30. Thus, the computational complexity of this case is $NP \times G = 900$. The computational complexity of the exhaustive search is $W^{M-1} = 32768$ while the PAPR for this case is 5.86dB. The PAPR by the iterative flipping algorithm for PTS (IPTS) [69] is 7.55dB with search complexity (M-1)W=30. The PAPR by the gradient descent method (GD) [71] with search complexity $C_{M-1}^{r} W^{r} I = C_{15}^{2} 2^{2} 3 = 1260$ is 6.96dB. The PAPR by ABC [10], PSO [72], and ACO [6], is 7.01dB, 7.13dB, and 6.52dB, respectively. Table 1 holds the comparison of the search complexity among the different methods for $CCDF = 10^{-3}$, NP=30, and G=30. It is obvious that ACO presents the better performance among the other methods with the same search complexity.

Method	Computational Complexity	PAPR (dB)
Original	0	10.84
Exhaustive	$W^{M-1} = 32768$	5.86
IPTS	(M-1)W=30	7.55
GD	$C_{M-1}^{r} W^{r} I = C_{15}^{2} 2^{2} 3 = 1260$	6.96
PSO	$NP \times G = 900$	7.13
ABC	$NP \times G = 900$	7.01
ACO	$NP \times G = 900$	6.52

Table 1. Comparison of computational complexity for CCDF=1e-3 among different PTS Schemes

4.3. Dual-band microwave filter design using SADE

Microwave filters are among the important components of a modern wireless communication system. Several papers exist in the literature that address the filter design problem [73]. Open

Loop Ring Resonator (OLRR) filters, which consist of two uniform microstrip lines and pairs of open loops between them, are widely used as the building block in several multiband bandpass filter design cases [74]. In [74], two pairs of folded OLRRs operating at two passbands are proposed to produce dual-band response.

A dual band OLRR filter is shown in Figure 4. The frequency response of such a filter depends on the filter dimensions and spacings between microstrip lines [74, 75]. The design parameters for this case are the ones shown in Figure 4, (W_1, W_2, L_1, L_2, L_3, L_4, L_5, S_1, S_2, S_3, G), all expressed in mm.

Such a filter design problem can be defined by the minimization of $|S_{11}|$ in the passband frequency range. This design problem is therefore defined by the minimization of the objective function:

$$F(\bar{x}) = 20\log\left\{\max\left|S_{11}(\bar{x}, f)\right|,\ f \in S_p\right\} \tag{30}$$

where $\bar{x}$ is the vector of filter geometry parameters and S_p is the set of distinct frequencies in the desired passband frequency ranges.

The filter is designed for operation in two WiMax (IEEE 802.16) frequency bands. These are the 3.5GHz and the 5.8GHz frequency bands. For this case, we set S_p={3.55, 3.6, 5.75, 5.8}. Figure 5 shows the simulated frequency response of this design. The simulated current distribution for the 3.6GHz and 5.8GHz frequencies is presented in Figure 6, where the resonating ring in each case is clearly seen. In the first passband between 3.508 and 3.809 GHz, the filter has a return loss less than 10dB and insertion loss greater than 0.5dB. In the second passband between 5.744 and 6.121 GHz the results also show a return loss less than 10dB and insertion loss greater than 0.5dB. The rejection band (between 4.236 and 5.367 GHz) has an insertion loss less than 20dB. In the first passband the return loss is less than 29dB at both 3.533 GHz and 3.759 GHz. In the second passband the return loss is less than 22dB at 5.794GHz and less than 28dB at 6.07GHz.

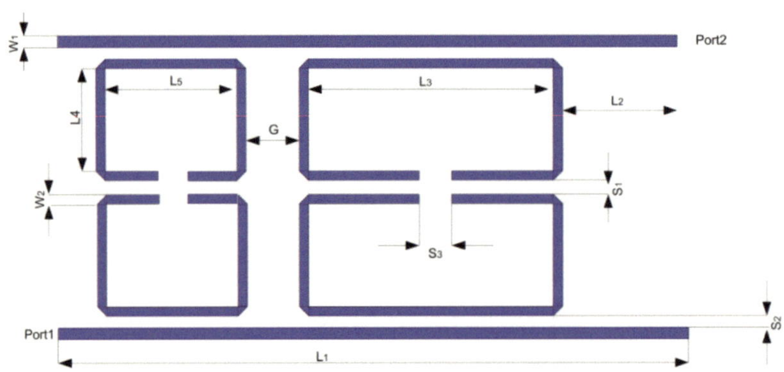

Figure 4. Dual-band filter geometry

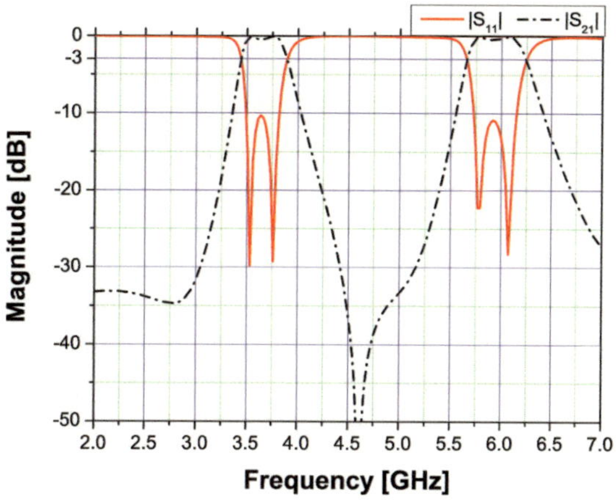

Figure 5. Simulated frequency response of the dual-band filter.

(a)

(b)

Figure 6. Current distribution simulations for dual-band filter at (a) 3.5GHz and (b) 5.8GHz

5. Conclusion

A brief survey of different evolutionary algorithms and their application to different problems in wireless communications has been presented. It must be pointed out that several evolutionary algorithms exist in the literature. GAs and SI algorithms are among those most commonly used. In order to select, the best algorithm for every problem one has to consider the problem characteristics. Another key issue is the selection of the algorithm control parameters, which is also in most cases problem-dependent. One may also use at first the control parameters for these algorithms that commonly perform well regardless of the characteristics of the problem to be solved. The example of the CSA problem in cellular networks showed the better performance of BB and BBExp compared to BPSO and ACO in terms of successfully obtained solutions and execution time. In the PTS optimization problem ACO outperformed ABC and PSO.

The selection of the SADE technique for microwave filter design has lead to a successful filter design which exhibits low loss in the passbands and high isolation between the passbands. The DE algorithms are also robust optimizers. In classical DE algorithms, the selection of the appropriate strategy for trial vector generation and control parameters requires additional computational time using a trial-and-error search procedure. Therefore, it is not always an easy task to fine-tune the control parameters and strategy given also that commonly the appropriate control parameters and strategy selection are problem dependent. The SaDE advantage though, is the fact that no additional time for solving a given problem is required. SaDE requires only the adjustment of two parameters: the population size and the number of iterations.

The practical examples subject to several constraints presented in this chapter show the applicability and the efficiency of using such algorithms.

Author details

Sotirios K. Goudos*

Address all correspondence to: sgoudo@physics.auth.gr

Radiocommunications Laboratory, Section of Applied and Environmental Physics, Department of Physics, Aristotle University of Thessaloniki, Thessaloniki, Greece

References

[1] Goldberg D. E. Genetic algorithms in search, optimization and machine learning. New York:Addison Wesley;1989.

[2] Holland J. H. Adaptation in natural and artificial systems. Ann Arbor: The University of Michigan Press;1975.

[3] Fogel D. B. Evolutionary computation: Toward a new philosophy of machine intelligence. IEEE Press, Piscataway, NJ;1995.

[4] Beyer H. G. and Schwefel H. P. Evolution strategies: A comprehensive introduction. Natural Computing 2002;1 (1) 3-52.

[5] Kennedy J. and Eberhart R., "Particle swarm optimization", in IEEE International Conference on Neural Networks, pp. 1942-1948, Piscataway, NJ, 1995.

[6] Dorigo M. and Stutzle T. Ant colony optimization. Cambridge, MA:The MIT Press; 2004.

[7] Karaboga D. and Basturk B. A powerful and efficient algorithm for numerical function optimization: Artificial bee colony (abc) algorithm. Journal of Global Optimization 2007;39 (3) 459-471.

[8] Clerc M., "The swarm and the queen: Towards a deterministic and adaptive particle swarm optimization", in Proceedings of the 1999 Congress on Evolutionary Computation, 1999. CEC 99 pp. 1951-1957, Washington, DC, 1999.

[9] Goudos S. K., Baltzis K. B., Bachtsevanidis C. and Sahalos J. N. Cell-to-switch assignment in cellular networks using barebones particle swarm optimization. IEICE Electron. Express 2010;7 (4) 254-260.

[10] Wang Y., Chen W. and Tellambura C. A papr reduction method based on artificial bee colony algorithm for ofdm signals. IEEE Transactions on Wireless Communications 2010;9 (10) 2994-2999.

[11] Zhu G. and Kwong S. Gbest-guided artificial bee colony algorithm for numerical function optimization. Applied Mathematics and Computation 2010;217 (7) 3166-3173.

[12] Dorigo M., Maniezzo V. and Colorni A. Ant system: Optimization by a colony of cooperating agents. IEEE Transactions on Systems, Man, and Cybernetics, Part B: Cybernetics 1996;26 (1) 29-41.

[13] Storn R. and Price K., "Differential evolution—a simple and efficient adaptive scheme for global optimization over continuous spaces", in Tech. Rep. TR-95-012, Berkeley, CA, 1995.

[14] Storn R. and Price K. Differential evolution-a simple and efficient heuristic for global optimization over continuous spaces. Journal of Global Optimization 1997;11 (4) 341-359.

[15] Qin A. K. and Suganthan P. N., "Self-adaptive differential evolution algorithm for numerical optimization", in The 2005 IEEE Congress on Evolutionary Computation, 2005., ed. P. N. Suganthan, pp. 1785-1791, 2005.

[16] Huang V. L., Qin A. K. and Suganthan P. N., "Self-adaptive differential evolution algorithm for constrained real-parameter optimization", in IEEE Congress on Evolutionary Computation, 2006. CEC 2006., ed. A. K. Qin, pp. 17-24, 2006.

[17] Qin A. K., Huang V. L. and Suganthan P. N. Differential evolution algorithm with strategy adaptation for global numerical optimization. IEEE Transactions on Evolutionary Computation 2009;13 (2) 398-417.

[18] Goudos S. K., Zaharis Z. D. and Yioultsis T. V. Application of a differential evolution algorithm with strategy adaptation to the design of multi-band microwave filters for wireless communications. Progress in Electromagnetics Research 2010;109 (123-137.

[19] Goudos S. K., Siakavara K., Samaras T., Vafiadis E. E. and Sahalos J. N. Sparse linear array synthesis with multiple constraints using differential evolution with strategy adaptation. IEEE Antennas and Wireless Propagation Letters 2011;10 (670-673.

[20] Kennedy J. Bare bones particle swarms. Proceedings of the IEEE Swarm Intelligence Symposium 2003 (SIS 2003) 2003;80-87.

[21] Pan F., Hu X., Eberhart R. and Chen Y., "An analysis of bare bones particle swarm", 2008.

[22] Dorigo M. and Gambardella L. M. Ant colonies for the travelling salesman problem. BioSystems 1997;43 (2) 73-81.

[23] Storn R. Differential evolution research-trends and open questions. Studies in Computational Intelligence 2008;143 (1-31.

[24] Mezura-Montes E., Velazquez-Reyes J. and Coello Coello C. A., "A comparative study of differential evolution variants for global optimization", in GECCO 2006-Genetic and Evolutionary Computation Conference, pp. 485-492, Seattle, WA, 2006.

[25] Iorio A. W. and Li X. Solving rotated multi-objective optimization problems using differential evolution. Lecture Notes in Artificial Intelligence (Subseries of Lecture Notes in Computer Science) 2004;3339 (861-872.

[26] Lee B. K., Chen H. W. and Chen B. S. Power control of cellular radio systems via robust smith prediction filter. IEEE Transactions on Wireless Communications 2004;3 (5) 1822-1831.

[27] Karamalis P. D., Skentos N. D. and Kanatas A. G. Selecting array configurations for mimo systems: An evolutionary computation approach. IEEE Transactions on Wireless Communications 2004;3 (6) 1994-1998.

[28] Lu H. Y. and Fang W. H. Joint receive antenna selection and symbol detection for mimo systems: A heterogeneous genetic approach. IEEE Communications Letters 2009;13 (2) 97-99.

[29] Lain J. K. Joint transmit/receive antenna selection for mimo systems: A real-valued genetic approach. IEEE Communications Letters 2011;15 (1) 58-60.

[30] Cheng W., Shi H., Yin X. and Li D. An elitism strategy based genetic algorithm for streaming pattern discovery in wireless sensor networks. IEEE Communications Letters 2011;15 (4) 419-421.

[31] Zheng G., Chatzinotas S. and Ottersten B. Generic optimization of linear precoding in multibeam satellite systems. IEEE Transactions on Wireless Communications 2012;11 (6) 2308-2320.

[32] Lu T. and Zhu J. Genetic algorithm for energy-efficient qos multicast routing. IEEE Communications Letters 2013;17 (1) 31-34.

[33] Wimalajeewa T. and Jayaweera S. K. Optimal power scheduling for correlated data fusion in wireless sensor networks via constrained pso. IEEE Transactions on Wireless Communications 2008;7 (9) 3608-3618.

[34] Yao W., Chen S., Tan S. and Hanzo L. Minimum bit error rate multiuser transmission designs using particle swarm optimisation. IEEE Transactions on Wireless Communications 2009;8 (10) 5012-5017.

[35] Kulkarni R. V. and Venayagamoorthy G. K. Particle swarm optimization in wireless-sensor networks: A brief survey. IEEE Trans Syst Man Cybern Pt C Appl Rev 2011;41 (2) 262-267.

[36] Nguyen H. A., Guo H. and Low K. S. Real-time estimation of sensor node's position using particle swarm optimization with log-barrier constraint. IEEE Trans. Instrum. Meas. 2011;60 (11) 3619-3628.

[37] Abdelhafiz A. H., Hammi O., Zerguine A., Al-Awami A. T. and Ghannouchi F. M. A pso based memory polynomial predistorter with embedded dimension estimation. IEEE Transactions on Broadcasting 2013;59 (4) 665-673.

[38] Li M., Xiong W. and Liang Q. Wireless sensor networks node localization algorithm based on improved abc algorithm. Chin. J. Sens. Actuators 2013;26 (2) 241-245.

[39] Li M., Xiong W. and Liang Q., "An improved abc-based node localization algorithm for wireless sensor network", in 2012 8th International Conference on Wireless Communications, Networking and Mobile Computing, WiCOM 2012, Shanghai, 2012.

[40] He P. Y. and Jiang M. Y., "Dynamic deployment of wireless sensor networks by an improved artificial bee colony algorithm", in 2013 International Conference on Sensors, Mechatronics and Automation, ICSMA 2013, pp. 862-866, Shenzhen, 2014.

[41] Gu Y., Xu X., Du J., Hou R. and Qian H. Anycast routing protocol for wireless sensor networks based on artificial bee colony. Chin. J. Sens. Actuators 2013;26 (4) 564-569.

[42] Berrocal-Plaza V., Vega-Rodríguez M. A., Gómez-Pulido J. A. and Sánchez-Pérez J. M., "Artificial bee colony algorithm applied to wimax network planning problem", in 2011 11th International Conference on Intelligent Systems Design and Applications, ISDA'11, pp. 504-509, Cordoba, 2011.

[43] Liu J., Jia Z., Qin X., Chang C., Xu G. and Xia X., "The applications in channel assignment based on cooperative hybrid artificial bee colony algorithm", in 2011 International Conference on Electrical Engineering and Automation, EEA 2011, pp. 401-406, Beijing, 2012.

[44] Deepalakshmi P. and Radhakrishnan S. An ant colony-based, receiver-initiated multicast mesh protocol for collaborative applications of mobile ad hoc networks. Eur Trans Telecommun 2014;25 (3) 354-369.

[45] Balaji V. and Duraisamy V. Ant optimized link quality for ad hoc on demand distance vector. Wireless Pers Commun 2014;

[46] Shirkande S. D. and Vatti R. A., "Aco based routing algorithms for ad-hoc network (wsn,manets):A survey", in 3rd International Conference on Communication Systems and Network Technologies, CSNT 2013, pp. 230-235, Gwalior, 2013.

[47] Liu X. and He D. Ant colony optimization with greedy migration mechanism for node deployment in wireless sensor networks. J Network Comput Appl 2014;39 (1) 310-318.

[48] Liu X. A transmission scheme for wireless sensor networks using ant colony optimization with unconventional characteristics. IEEE Communications Letters 2014;18 (7) 1214-1217.

[49] Liu W. and Wang L. Ant colony optimization routing algorithm based on wsn. ICIC Express Lett Part B Appl. 2013;4 (3) 541-547.

[50] Li Z. and Shi Q., "An qos algorithm based on aco for wireless sensor network", in 15th IEEE International Conference on High Performance Computing and Communications, HPCC 2013 and 11th IEEE/IFIP International Conference on Embedded and Ubiquitous Computing, EUC 2013, pp. 1671-1674, IEEE Computer Society, Zhangjiajie, Hunan, 2014.

[51] Liu X. Sensor deployment of wireless sensor networks based on ant colony optimization with three classes of ant transitions. IEEE Communications Letters 2012;16 (10) 1604-1607.

[52] He Q., Feng Z. and Zhang P. Reconfiguration decision making based on ant colony optimization in cognitive radio network. Wireless Pers Commun 2013;71 (2) 1247-1269.

[53] Zhao Y., Xu X., Hao Z., Tao X. and Zhang P., "Resource allocation in multiuser ofdm system based on ant colony optimization", in IEEE Wireless Communications and Networking Conference 2010, WCNC 2010, Sydney, NSW, 2010.

[54] Marinello J. C. and Abrao T., "Lattice reduction aided detector for dense mimo via ant colony optimization", in 2013 IEEE Wireless Communications and Networking Conference, WCNC 2013, pp. 2839-2844, Shanghai, 2013.

[55] Lain J. K. and Chen J. Y. Near-mld mimo detection based on a modified ant colony optimization. IEEE Communications Letters 2010;14 (8) 722-724.

[56] Das S., Mukherjee R., Kundu R. and Vasilakos T., "Multi-user detection in multi-carrier cdma wireless broadband system using a binary adaptive differential evolution

algorithm", in 2013 15th Genetic and Evolutionary Computation Conference, GECCO 2013, pp. 1245-1252, Amsterdam, 2013.

[57] Kundu S., Das S., Vasilakos A. V. and Biswas S. A modified differential evolution-based combined routing and sleep scheduling scheme for lifetime maximization of wireless sensor networks. Soft Computing 2014;

[58] Yin X., Ling Z., Guan L. and Liang F. Minimum distance clustering algorithm based on an improved differential evolution. Int. J. Sens. Netw. 2014;15 (1) 1-10.

[59] Wu C. Y., Yen H. S., Chiu C. C., Chiang J. S. and Chang C. W., "Urban area propagation path loss reduction by dynamic differential evolution algorithm", in 1st International Conference on Intelligent Green Building and Smart Grid, IGBSG 2014, IEEE Computer Society, Taipei, 2014.

[60] Wang Z. and Zhang W. Spectrum sharing with limited channel feedback. IEEE Transactions on Wireless Communications 2013;12 (5) 2524-2532.

[61] Li K., Wang X. and Ping L. Analysis and optimization of interleave-division multiple-access communication systems. IEEE Transactions on Wireless Communications 2007;6 (5) 1973-1982.

[62] Zhao Z., Peng Z., Zheng S. and Shang J. Cognitive radio spectrum allocation using evolutionary algorithms. IEEE Transactions on Wireless Communications 2009;8 (9) 4421-4425.

[63] Merchant A. and Sengupta B. Assignment of cells to switches in pcs networks. IEEE/ACM Transactions on Networking 1995;3 (5) 521-526.

[64] Pierre S. and Houéto F. A tabu search approach for assigning cells to switches in cellular mobile networks. Computer Communications 2002;25 (5) 464-477.

[65] Shyu S. J., Lin B. M. T. and Hsiao T. S. Ant colony optimization for the cell assignment problem in pcs networks. Computers and Operations Research 2006;33 (6) 1713-1740.

[66] Salcedo-Sanz S. and Yao X. Assignment of cells to switches in a cellular mobile network using a hybrid hopfield network-genetic algorithm approach. Appl. Soft Comput. J. 2008;8 (1) 216-224.

[67] Kennedy J. and Eberhart R. C., "Discrete binary version of the particle swarm algorithm", in Proceedings of the IEEE International Conference on Systems, Man and Cybernetics, pp. 4104-4108, 1997.

[68] Müller S. H. and Huber J. B. Ofdm with reduced peak-to-average power ratio by optimum combination of partial transmit sequences. Electronics Letters 1997;33 (5) 368-369.

[69] Cimini Jr L. J. and Sollenberger N. R. Peak-to-average power ratio reduction of an ofdm signal using partial transmit sequences. IEEE Communications Letters 2000;4 (3) 86-88.

[70] Tellambura C. Improved phase factor computation for the par reduction of an ofdm signal using pts. IEEE Communications Letters 2001;5 (4) 135-137.

[71] Han S. H. and Lee J. H. Papr reduction of ofdm signals using a reduced complexity pts technique. IEEE Signal Processing Letters 2004;11 (11) 887-890.

[72] Wen J. H., Lee S. H., Huang Y. F. and Hung H. L. A suboptimal pts algorithm based on paticle swarm optimization for papr reduction in ofdm systems. EURASIP J. Wireless Commun. Networking 2008;2008 (14)

[73] Wang X. H., Wang B. Z. and Chen K. J. Compact broadband dual-band bandpass filters using slotted ground structures. Progress in Electromagnetics Research 2008;82 (151-166.

[74] Chen C. Y. and Hsu C. Y. A simple and effective method for microstrip dual-band filters design. IEEE Microwave and Wireless Components Letters 2006;16 (5) 246-248.

[75] Koziel S. and Bandler J. W. Space mapping with multiple coarse models for optimization of microwave components. IEEE Microwave and Wireless Components Letters 2008;18 (1) 1-3.

Chapter 2

User Selection and Precoding Techniques for Rate Maximization in Broadcast MISO Systems

E. Castañeda, A. Silva and A. Gameiro

Additional information is available at the end of the chapter

http://dx.doi.org/10.5772/58937

1. Introduction

For broadcast channels (BC) in multiuser multiple-input multiple-output (MU-MIMO) systems, the overall achievable performance depends on the efficient use of the resources at the base station (BS). When channel state information (CSI) is known at the transmitter the gain is twofold: full spatial degrees of freedom can be attained [1] and the system can be optimized over a new degree of freedom given by the users. In the literature of MU-MIMO systems two dimensions of optimization are studied: *multiplexing diversity* and *multiuser diversity* (MUD). The former is a consequence of the independent fading across all MIMO links, which yields a set of parallel spatial channels where different data streams can be transmitted increasing the system capacity [2]. The latter is given when users that are geographically far apart have channels that fade independently at any point in time. For the system sum rate maximization, such independent fading process is exploited so that the user with the largest channel magnitude will be selected for transmission. In MU-MIMO systems spatial and temporal communication can be performed in two ways: 1) orthogonal multiple access (time, frequency, or code) and 2) spatial-division multiple access (SDMA). To achieve orthogonality in time, all the spatial resources at the BS (N_t antennas) are used to communicate with one user (N_r antennas) at a time, which is known also as time-division multiple access (TDMA). This technique avoids inter-user interference, achieves power gains that scale with N_t, enhances data rates for a single user specially at low signal-to-noise ratio (SNR) regime, and is robust to CSI uncertainty [1], [3]. SDMA exploits CSI at the BS allowing $K>1$ users to be scheduled at the same time achieving multiplexing gains of $\min(N_t, KN_r)$ at high SNR regime, where the system is limited by the degrees of freedom and not by power. Since the BS has knowledge of the data symbols and CSI, the multiuser transmission can be optimized using coding techniques. The achievable rate region in BC is based on dirty paper coding (DPC) [4].

The principle behind this optimum coding technique is that the BS knows the interference for a given user and can pre-subtract it before transmission, which yields the capacity of an interference free channel. However, DPC is a nonlinear process that requires successive encoding and decoding whose complexity is prohibitive in practical systems and other suboptimal techniques are preferred instead. In the literature, DPC has been interpreted as *beamforming* (BF) [5] which is a SDMA scheme where data streams of different users are encoded independently and multiplied by weight vectors in order to mitigate mutual interference. Although BF is a suboptimal multiuser transmission scheme, several works (e.g., [3], [5]) have shown that it can achieve a large portion of the DPC rate and its performance is close-to-optimal for large N_t and K. Nevertheless, the optimization of the downlink BF weight vectors is a non-convex problem [6]. The specific selection of the weight vector of a given user may affect the performance of other users, i.e., the achievable signal-to-interference-plus-noise ratio (SINR) of one user is coupled with the other users weight vectors and transmit powers [6]. Since the weight vectors must be optimized jointly, optimum BF is a complicated task and suboptimal weights given by Zero-Forcing (ZF) based methods can be used [5]. The joint optimization of the beamforming weights, the power allocation, and the set of links that are scheduled under SDMA mode is performed for a given global objective function. In BC system literature a meaningful figure of merit is the total sum rate since it quantifies the total data flow for a specific BF scheme and the open problems are power allocation and user selection. If the set of scheduled users meets that $N_t \geq K N_r$, power allocation can be given either by Lagrangian methods (convex optimization [7]) or by equal power allocation (close-to-optimal for the high SNR regime [8]–[12]). However, when the set of users is larger than the number of spatial resources at the BS, i.e., $N_t < K N_r$, user selection is required. The selection of the optimum set of users that maximizes the sum rate for a given BF scheme with optimum power allocation is a mixed non-convex problem. Indeed, in systems where users must be allocated in different radio resources (slots or sub-channels) finding the optimum subsets of users under SDMA mode for each radio resource is a NP-complete problem whose optimal solution can only be found via exhaustive search (ES) [13]. Recent research works have proposed a number of feasible heuristics algorithms that find a suboptimal yet acceptable solution to the sum rate maximization problem with SDMA communication. The literature of MU-MIMO has been focused on ZF-based BF schemes due to their tractability and the fact that some properties of the wireless channels can be used to estimate the reliability of joint transmission for a given set of users. The main objective of joint scheduling and BF is to make better decisions at the media access control (MAC) layer by exploiting information from the physical (PHY) layer. The literature reviewed in this chapter addresses the scheduling (user selection) process in the MAC layer using PHY layer information without considering constraints from upper layers. Our aim is to provide a comprehensive overview of algorithms proposed over the last years regarding joint user selection and SDMA schemes that solve the sum rate maximization problem in MU-MISO BC systems.

2. System model

Consider a single-cell with a single base station equipped with N_t antennas, and a set S of K single-antenna users ($N_r=1$) illustrated in Fig. 1. We assume perfect CSI at the BS and the channel coefficients are modeled as independent random variables, with a zero-mean circularly symmetric complex Gaussian distribution (Rayleigh fading). The signal received by the kth user is given by:

$$y_k = \sqrt{p_k}\mathbf{h}_k\mathbf{w}_k s_k + \sum_{i \neq k}^{K}\sqrt{p_i}\mathbf{h}_k w_i s_i + n_k = \mathbf{h}_k x + n_k \tag{1}$$

where s_k is the intended symbol for user k, $\mathbf{x}\in\mathbb{C}^{N_t\times 1}$ is the transmitted signal vector from the base station antennas, $\mathbf{h}_k\in\mathbb{C}^{1\times N_t}$ is the channel vector to the user k. Each user ignores the modulation and coding of other users, i.e., it is assumed single-user detection where each user treats the signals intended for other users as interference. $n_k \backsim \mathcal{CN}(0, \sigma_n^2)$ is the additive zero mean white Gaussian noise with variance σ_n^2. The entries of the block fading channel $\mathbf{H}=[\mathbf{h}_1^T, \cdots, \mathbf{h}_K^T]^T$ are normalized so that they have unitary variance, and the transmitted signal $\mathbf{x}=\sum_{k=1}^{K}\sqrt{p_k}\mathbf{w}_k s_k$ has an average power constraint $\mathbb{E}[\mathbf{x}^H\mathbf{x}]\leq P$ where $\mathbb{E}[\bullet]$ is the expectation operation. Since the noise has unit variance, P represents the SNR.

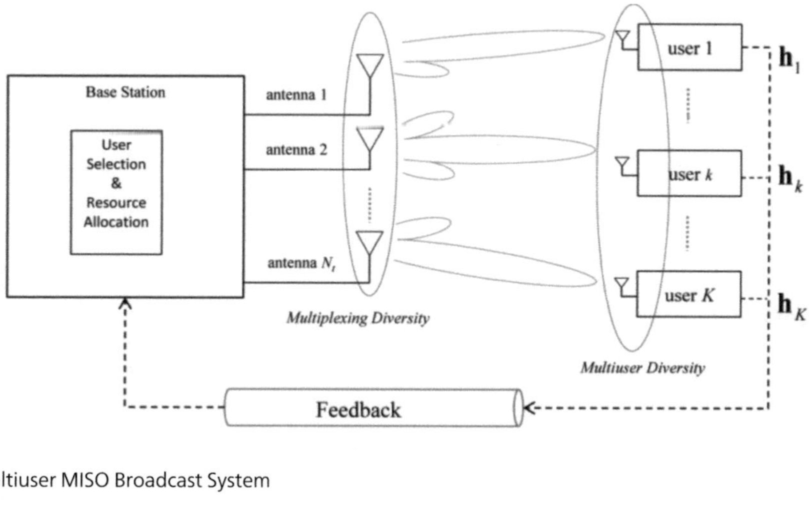

Figure 1. Multiuser MISO Broadcast System

For linear spatial processing at the transmitter, the BF matrix can be defined as $\mathbf{W} = [\mathbf{w}_1, \cdots, \mathbf{w}_K]$, the symbol vector as $\mathbf{s}=[s_1, \cdots, s_K]$ and $\mathbf{P}=diag[p_1, \cdots, p_K]$ is the matrix whose main diagonal contains the powers. The SINR of the kth user is given by:

$$\mathrm{SINR}_k = \frac{p_k \left|\mathbf{h}_k \mathbf{w}_k\right|^2}{\sum_{i \neq k} p_i \left|\mathbf{h}_k \mathbf{w}_i\right|^2 + \sigma_n^2} \tag{2}$$

and the instantaneous achievable data rate of user k is $r_k = \log_2(1 + \mathrm{SINR}_k)$.

3. Problem formulation

The performance of a MIMO system is measured by a global objective function of the individual data rates or SINRs $U(r_1, \cdots, r_K)$. From the system perspective it is desirable to optimize $U(\bullet)$ instead of the individual rates $r_i \ \forall i \in S$ since the latter are coupled by the transmit powers and the beamforming weights in (2). Thus the performance depends on how efficiently the resources are allocated to each user and how effectively the interference from other users is mitigated. In this chapter we optimize the global utility function modeled as the sum rate maximization problem using BF subject to global power constraints. For the case where $K \leq N_t$ the general optimization problem is given by

$$\mathrm{R}^{BF} = \max_{\mathbf{W},\mathbf{P}} \sum_{k=1}^{K} \beta_k r_k \quad \text{s.t.} \quad \left\{ \left\| \mathbf{W}\mathbf{P}^{1/2} \right\|_F^2 \leq P \tag{3}$$

where $\| \bullet \|_F$ denotes the Frobenius norm, β_k is a priority weight associated to user k defined a priory by upper layers of the communications system to take into account QoS, fairness, or another system constraint. Finding a solution of (3) is a complex problem due to the nature of the optimum **P** and **W** and each solution depends on the system requirements expressed by the weights β_k [14]. The computation of optimal beamforming weights $\mathbf{w}_k$ involves SINR balancing [11] and since the weights do not have a closed-form, iterative computational demanding algorithms have been proposed to determine them [6], [15]. Indeed, problem (3) is NP-hard even when all priority weights β_k are equal [16].

3.1. Multiuser scenario

Let $\Omega = \{1, \cdots, K\}$ be the set of all competing users where K is larger than the number of available antennas at the BS, i.e., $|\Omega| = K \geq N_t$, where $|\Omega|$ denotes the cardinality of the set Ω. In order to exploit the optimization dimension provided by MUD, it is necessary to select a set of users S whose channel characteristics maximize the sum rate when they transmit simultaneously in the same radio resource. Such characteristics are defined by the type of beamforming scheme, the power constraints, the SNR regime, and the deployment characteristics (N_t and K). The sum rate maximization with user selection optimization problem can be defined as:

$$\Re(\mathrm{S}) = \max_{\mathrm{S}\subset\Omega: |\mathrm{S}|\leq N_t} R^{BF}(\mathbf{H}(\mathrm{S})) \tag{4}$$

where $\mathbf{H}(S)$ is the row-reduced channel matrix containing only the channels of the subset of selected users S and $R^{BF}(\mathbf{H}(S))$ is the achievable sum rate for such set. User scheduling is a real time operation whose computational complexity and implementation efficiency affect the performance of upper-layers. Moreover, finding the optimal solution of (4) requires an exhaustive search over a search space of size $L = \sum_{i=1}^{N_t} K!/(i!(K-i)!)$, which is the number of all ordered combinations of users.

Since the computation of the optimal solution of the sum rate maximization problem implies the joint optimization over **P**, **W**, and S, the original problem (4) can be relaxed by taking one or more of the following actions: 1) by using beamforming weights with a defined structure; 2) based on linear beamforming the power allocation can be performed either by Lagrangian methods or by equal power allocation; and 3) based on the structure of the linear beamforming it is possible to design user selection algorithms that exploit information contained in **H** and at the same time, to achieve a good trade-off between performance and complexity. In the following sections we will study systems with equally prioritized users ($\beta_k = 1\ \forall k \in \Omega$) and the characteristics of state-of-the-art user selection algorithms that find suboptimal yet acceptable solutions to the non-convex and combinatorial problem (4).

4. Linear beamforming

In this section we describe the structure of two sub-optimal linear beamforming schemes and the optimal power allocation for each one of them. It is assumed perfect CSI at the transmitter which can be attained through time-division duplex (TDD) scheme assuming channel reciprocity [6]. Notice that the weight vectors multiply the intended symbols in (1) which can be seen as a form of precoding, henceforward we use the terms beamforming and precoding interchangeably.

4.1. Zero Forcing Beamforming

In Zero Forcing Beamforming (ZFBF), the channel matrix **H** at the transmitter is processed so that orthogonal channels between the transmitter and the receiver are created, defining a set of parallel subchannels [5]. The Moore-Penrose pseudo inverse of $\mathbf{H}(S)$ is given by [17]:

$$\bar{\mathbf{W}}(\mathrm{S}) = \mathbf{H}(\mathrm{S})^H(\mathbf{H}(\mathrm{S})\mathbf{H}(\mathrm{S})^H)^{-1} \tag{5}$$

and the ZFBF matrix is given by the normalized column vectors of (5) as $\mathbf{W}(S) = [\bar{\mathbf{w}}_1/\|\bar{\mathbf{w}}_1\|, \cdots, \bar{\mathbf{w}}_{|S|}/\|\bar{\mathbf{w}}_{|S|}\|]$. Under ZFBF scheme the sum rate maximizing power allocation is given by the water-filling algorithm and according to [5] the information rate achieved with optimum **P** in (3) is given by :

$$R^{ZFBF}(\mathbf{H}(S)) = \sum_{i=1}^{|S|} \left(\log(\mu b_i)\right)^+ \tag{6}$$

where $b_i = \left([\mathbf{H}(S)\mathbf{H}(S)^{\mathrm{H}}]_{ii}^{-1}\right)^{-1}$ is the effective channel gain of the ith user and its allocated power is $p_i = (\mu b_i - 1)^+$, the water level μ is chosen to satisfy $\sum_{i\in S} \left(\mu - b_i^{-1}\right)^+ = P$ and $(x)^+ = \max\{x, 0\}$. If all users in S are allocated with nonzero power, the water level has the compact form [10], [18]: $\mu = \frac{1}{|S|}\left(\|\overline{\mathbf{W}}(S)\|_F^2 + P\right) = \frac{1}{|S|}\left(P + \sum_{i\in S} b_i^{-1}\right)$.

4.2. Zero forcing dirty paper

Suboptimal throughput maximization in Gaussian BC channels has been proposed in several works [5], [18], [19] based on the QR-type decomposition of the channel matrix $\mathbf{H}(S) = \mathbf{L}(S)\mathbf{Q}(S)$ obtained by applying Gram-Schmidt orthogonalization (GSO) [20]. $\mathbf{L}(S)$ is a lower triangular matrix and $\mathbf{Q}(S)$ has orthonormal rows. The beamforming matrix given by $\mathbf{W}(S) = \mathbf{Q}(S)^{\mathrm{H}}$ generates a set of interference channels $y_i = l_{ii}\sqrt{p_i}s_i + \sum_{j<i} l_{ij}\sqrt{p_j}s_j + n_i$, $i = 1, \cdots, k$ while no information is sent to users $k+1, \cdots, K$. In order to eliminate the interference component $I_i = \sum_{j<i} l_{ij}\sqrt{p_j}s_j$ of the ith user, the signals $\sqrt{p_j}s_j$ for $i = 1, \cdots, k$ are obtained by successive dirty paper encoding, where I_i is non-causally known. This precoding scheme was proposed in [5] and the authors showed that the precoding matrix forces to zero the interference caused by users $j > i$ on each user i, therefore this scheme is called zero-forcing dirty-paper (ZFDP) coding. The information rate achieved with optimum **P** in (3) under the ZFDP scheme is given by [5]:

$$R^{ZFDP}(\mathbf{H}(S)) = \sum_{i=1}^{|S|} \left(\log(\mu d_i)\right)^+ \tag{7}$$

where $d_i = |l_{ii}|^2$ is the squared absolute value of l_{ii} and μ is the solution to the water-filling equation $\sum_{i\in S} \left(\mu - d_i^{-1}\right)^+ = P$, which defines the ith power as $p_i = (\mu d_i - 1)^+$.

5. Metrics of spatial compatibility

The sum rate maximization problem (4) can be solved by fixing the precoder structure and power allocation method. Under ZF-based precoding the performance strongly depends on the spatial correlation between the components of $\mathbf{H}(S)$. The more correlated the channels, the higher the power penalty imposed by ZF schemes which yields a degradation of the achievable SINRs and a poor system performance. For this reason, problem (4) has been tackled in the literature by optimizing the spatial compatibility between scheduled users. This is accomplished by optimizing a specific metric over the channel matrix which can be related or can provide indirect information of the achievable sum rate for a given set of users and channel realization. A metric of spatial compatibility is a function of the CSI at the transmitter so that $f:\mathbb{C}^{|S|\times N_t}\mapsto\mathbb{R}_+$ where the mapping quantifies how profitable is to select S for transmission. Different metrics for spatial compatibility have been proposed in the literature and this section presents a unified treatment of the most common metrics used by several algorithms that solve (4) sub-optimally. It is worth mentioning that the optimization over a given metric may bring some advantages in terms of computational complexity, for instance, iterative evaluation of $f(\mathbf{H}(S))$ does not require the computation of the optimum power allocation. Some metrics are given by simple relations between the row vectors in $\mathbf{H}(S)$ which avoids matrix operations. Under certain SNR constraints described in Section 6, the user set that solves problem (4) achieves maximum multiplexing diversity, i.e., its cardinality is equal to N_t [5]. In such a SNR regime, the search space of the solution of problem (4) is reduced from L to $L_{N_t}=K!/(N_t!(K-N_t)!)$ and optimization over $f(\mathbf{H}(S))$ can simplify the search of the optimum user set S. In this section we introduce the channel metrics, its main properties, and the relations between them.

5.1. Null space projection

Considering a given set of users S with channels $\mathbf{h}_i\,\forall\, i\in S$ and for ZFBF the effective channel gain of the ith selected user defined in Section 4.1 is given by [5]:

$$b_i=\frac{1}{[(\mathbf{H}(S)\mathbf{H}(S)^H)^{-1}]_{i,i}}=\left\|\mathbf{h}_i\mathbf{Q}_{\mathcal{V}_i}\right\|^2=\left\|\mathbf{h}_i\right\|^2\sin^2\theta_{\mathcal{V}_i\mathbf{h}_i} \tag{8}$$

where $\mathbf{P}_{\mathcal{V}_i}=\widetilde{\mathbf{H}}_i^H(\widetilde{\mathbf{H}}_i\widetilde{\mathbf{H}}_i^H)^{-1}\widetilde{\mathbf{H}}_i$ is the projector matrix onto $\mathcal{V}_i=Sp(\widetilde{\mathbf{H}}_i)$ the subspace spanned by the rows of the aggregate interference matrix $\widetilde{\mathbf{H}}_i=[\mathbf{h}_1^H,\ \cdots,\ \mathbf{h}_{i-1}^H,\ \mathbf{h}_{i+1}^H,\ \cdots,\ \mathbf{h}_{|S|}^H]^H$ associated with user i. $\mathbf{Q}_{\mathcal{V}_i}=\mathbf{I}-\mathbf{P}_{\mathcal{V}_i}$ is the projector matrix onto the null space of $\mathcal{V}_i$. The operation in (8) is equivalent to the projection of $\mathbf{h}_i$ onto the null space spanned by the channel components of $\widetilde{\mathbf{H}}_i$ illustrated in Fig. 2. Notice that $\|\mathbf{h}_i\|^2$ in (8) is affected by the weight $\sin^2\theta_{\mathcal{V}_i\mathbf{h}_i}$ which is the squared sine of the angle between the channel vector $\mathbf{h}_i$ and the subspace spanned by the components of $\widetilde{\mathbf{H}}_i$. The weight $\sin^2\theta_{\mathcal{V}_i\mathbf{h}_i}$ is referred in the literature as the *projection power loss factor* since it will affect the effective amount of power that is transmitted over the ith link. Using the properties of water-filling and the strong relationship between the sum rate maximization and the maximization of the terms b_i we elaborate a compact formulation of the maximization of metric (8). Theoretical results in [9] show that for the MISO BC system a meaningful metric to estimate the achievable performance of S is given by $\prod_{i\in S} b_i$ under certain constraints over the water level μ. However, for single antenna receivers the performance gap between the sum and the product of the terms b_i is negligible and hereafter we analyze the metric $\sum_{i\in S} b_i$ which is equivalent to a matrix trace operation. Let $\bar{\mathbf{H}}(S)=\mathbf{H}(S)\mathbf{H}(S)^H$ and $\dot{\mathbf{H}}(S)=\bar{\mathbf{H}}(S)^{-1}$ be a Wishart matrix and its respective inverse which characterize the interaction of all user channels in S. Considering the definitions in (6) and (8) the set of users that achieves a suboptimal solution to problem (4) maximizing the sum of the effective channel gains is given by the set of users that optimally solves of the following combinatorial problem:

$$S_\omega = \arg \min_{S\subset\Omega:|S|=N_t} \mathrm{tr}(\dot{\mathbf{H}}(S)) \tag{9}$$

where tr(•) denotes the trace operator. The optimum set S_ω for the NSP metric is unique and it will contain the users that maximize $\sum_{i\in S} \|\mathbf{h}_i\|^2 \sin^2 \theta_{V_i \mathbf{h}_i}$. The selection of S based on the NSP in (9) yields a close-to-optimal solution to problem (4) for a ZF-based precoding under the following conditions: $P>P_0$, $K>N_t$, or large values of both K and N_t. The term P_0 is a critical SNR value that depends on $\mathbf{H}(S)$ in order to meet the cardinality constraint over S [5]. The evaluation of the metric (8) is not unique and several algorithms described in next section use different operations to compute or approximate it in order to define algorithms that solve (9) with less computational complexity than the exhaustive search, especially when $K\to\infty$. Next we present several methods to evaluate the NSP for ZFBF and ZFDP. The application of each one of these methods lies in the complexity required to evaluate them and the available CSI at the BS. In other words, each method represents a trade-off between accuracy of the NSP evaluation and the required CSI at the transmitter.

5.1.1. Orthogonal projector for ZFBF

The computation of $\mathbf{Q}_{V_i}$ is not unique and different forms to evaluate such matrix can be efficiently used in different contexts, i.e., depending on the available CSI at the transmitter, the required computational complexity, and the deployment. Let us decompose matrix $\tilde{\mathbf{H}}_i$ of the ith user by means of singular value decomposition (SVD) [17] as follows:

$$\tilde{\mathbf{H}}_i = \mathbf{U}_{\tilde{\mathbf{H}}_i} \Sigma_{\tilde{\mathbf{H}}_i} [\bar{\mathbf{V}}_{\tilde{\mathbf{H}}_i} \mathbf{V}_{\tilde{\mathbf{H}}_i}]^H \tag{10}$$

where $\tilde{\mathbf{V}}_{\tilde{\mathbf{H}}_i}$ contains the N_t - r basis of the null space of $\tilde{\mathbf{H}}_i$ and $r=rank(\tilde{\mathbf{H}}_i)$.

The orthogonal projector matrix is given by $\mathbf{Q}_{\mathcal{V}_i} = \widetilde{\mathbf{V}}_{\tilde{\mathbf{H}}_i} \widetilde{\mathbf{V}}_{\tilde{\mathbf{H}}_i}{}^H$ and the set S_ω in (9) maximizes the objective function $\sum_i \| \mathbf{h}_i \mathbf{Q}_{\mathcal{V}_i} \|^2$. In some scenarios described in the following sections, it is assumed that the BS knows the basis of $\mathcal{V}_i$ for any user $i \in S$. Let $\{\mathbf{v}_j\}_{j=1}^r$ be the column vectors of $\overline{\mathbf{V}}_{\tilde{\mathbf{H}}_i}$ defined in (10) and the NSP in (8) for the ith user can be computed as follows:

$$\left\| \mathbf{h}_i \mathbf{Q}_{\mathcal{V}_i} \right\|^2 = \left\| \mathbf{h}_i \left(\mathbf{I} - \sum_{j=1}^{r} \mathbf{v}_j \mathbf{v}_j^H \right) \right\|^2 \tag{11}$$

The NSP operation in (11) can be also computed by applying GSO to $\mathbf{H}(S)$ as in [21] which represents a lower computational complexity than the SVD approach [22]. Using the basis of $\mathcal{V}_i$ the magnitude of the NSP operation is given by $\| \mathbf{h}_i \mathbf{Q}_{\mathcal{V}_i} \|^2 = \| \mathbf{h}_i \|^2 - \| \hat{\mathbf{h}}_i \|^2$ where $\hat{\mathbf{h}}_i$ is the projection of $\mathbf{h}_i$ onto each one of the orthogonal basis of $\mathcal{V}_i$ given by [20]:

$$\hat{\mathbf{h}}_i = \sum_{j=1}^{r} \frac{\left\| \mathbf{h}_i \right\| \cos \theta_{\mathbf{h}_i \mathbf{v}_j}}{\left\| \mathbf{v}_j \right\|} \mathbf{v}_j^H \tag{12}$$

where the term $\cos \theta_{\mathbf{h}_i \mathbf{v}_j}$ represents the coefficient of correlation between the vectors $\mathbf{h}_i$ and $\mathbf{v}_j$ defined as [17]:

$$\cos \theta_{\mathbf{h}_i \mathbf{v}_j} = \frac{\mathbf{h}_i \mathbf{v}_j^H}{\left\| \mathbf{h}_i \right\| \left\| \mathbf{v}_j \right\|} \tag{13}$$

The domain of the coefficient is $0 \le \eta_{\mathbf{h}_i \mathbf{v}_j} = \cos \theta_{\mathbf{h}_i \mathbf{v}_j} \le 1$ and $\theta_{\mathbf{h}_i \mathbf{v}_j} = \frac{\pi}{2}$ means perfect spatial orthogonality. The NSP computation is not unique and different matrix operations can be use to evaluate it. Using the full channel matrix $\mathbf{H}(S)$ and $\widetilde{\mathbf{H}}_i$ for all $i \in S$ the block matrix determinant formula to compute $\det(\mathbf{H}(S)\mathbf{H}(S)^H)$ reads [23]:

$$\det\left(\mathbf{H}(\mathrm{S}) \mathbf{H}(\mathrm{S})^H \right) = \det\left(\tilde{\mathbf{H}}_i \tilde{\mathbf{H}}_i^H \right) \left\| \mathbf{h}_i \mathbf{Q}_{\mathcal{V}_i} \right\|^2 \tag{14}$$

The orthogonal projection defined in (8) has a direct relationship with the correlation coefficients defined in (13). The normalized power loss experienced by the ith channel when it interacts with the subspace $\mathcal{V}_i$ is called the coefficient of determination given by [17]:

$$\Delta^2_{\mathrm{V}_i\mathbf{h}_i} = \frac{\mathbf{h}_i\mathbf{P}_{\mathrm{V}_i}\mathbf{h}_i^H}{\mathbf{h}_i\mathbf{h}_i^H} \tag{15}$$

where $\Delta^2_{\mathcal{V}_i\mathbf{h}_i}$ measures how much the vector $\mathbf{h}_i$ can be predicted from the channel vectors of $\widetilde{\mathbf{H}}_i$. Notice that from (8) and (15) the projection power loss factor of $\mathbf{h}_i$ due to the projection onto the null space of $\mathcal{V}_i$ is equivalent to $1 - \Delta^2_{\mathcal{V}_i\mathbf{h}_i}$ which can be evaluated as follows [17]:

$$1-\Delta^2_{\mathrm{V}_i\mathbf{h}_i} = (1-\eta^2_{\mathbf{h}_i\pi(1)})\ldots(1-\eta^2_{\mathbf{h}_i\pi(k)|\pi(1)\ldots\pi(k-1)}) \tag{16}$$

where $\pi(k)$ is the kth ordered element of $\widetilde{\mathbf{H}}_i$ and $\eta_{\mathbf{h}_i\pi(k)|\pi(1)\cdots\pi(k\text{-}1)}$ is the partial correlation between the channel vector $\mathbf{h}_i$ and the selected vector associated with $\pi(k)$ eliminating the effects due to $\pi(1)\pi(2)\cdots\pi(k - 1)$. Using multiple regression analysis it is possible to evaluate $\|\mathbf{h}_i\mathbf{Q}_{\mathcal{V}_i}\|^2 = \|\mathbf{h}_i\|^2(1 - \Delta^2_{\mathcal{V}_i\mathbf{h}_i})$ by extracting the partial correlation coefficients from the correlation coefficients choosing one user order π out of $(|S| - 1)!$ permutations of the users in S [17]. A different approach can be applied if for a given set of channels $\mathbf{h}_j$ $\forall j \in S$ the orthogonal projector matrix of each channel is known so that $\mathbf{Q}_j = \mathbf{I} - \mathbf{h}_j^H(\mathbf{h}_j\mathbf{h}_j^H)^{-1}\mathbf{h}_j$.

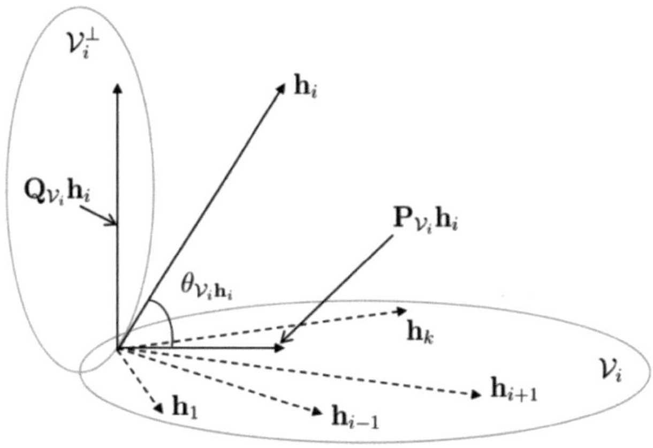

Figure 2. The spatial relationship between the components of vector $\mathbf{h}_i$ and $\mathcal{V}_i$.

From [24] we have the following result:

$$\mathbf{Q}_{\mathrm{V}_i} = \left(\prod_{j\neq i, j\in \mathrm{S}} \mathbf{Q}_j\right)^n, n \to \infty \tag{17}$$

which establishes that the orthogonal projector matrix onto $\mathcal{V}_i$ can be approximated by recurrently projecting onto independent orthogonal subspaces such that their intersection strongly converges to $\mathbf{Q}_{\mathcal{V}_i}$ as n grows. The NSP measured by $\mathrm{tr}(\dot{\mathbf{H}}(S))$ has been used by several user selection algorithms either by avoiding the exhaustive search required to solve (9) (e.g., [18], [19], [21]) or by using approximations of metric (8) (e.g., [25], [26]).

5.1.2. Orthogonal projector for ZFDP

In Section 4.2 the ZFDP beamforming was introduced and it was mentioned that the received signal of user i contains an interference component from all $j<i$ users, i.e., the previously encoded users. Given an specific encoding order $\pi(i)$, $i\in\{1, \cdots, |S|\}$ (a permutation of the users in S), the beamforming vectors $\mathbf{w}_{\pi(i)}$ are computed either by a QR-type decomposition or by GSO as follows [27]:

$$\mathbf{w}_{\pi(i)}=\frac{\mathrm{T}_i\mathbf{h}_{\pi(i)}^H}{\|\mathrm{T}_i\mathbf{h}_{\pi(i)}^H\|}, \qquad \mathrm{T}_i=\begin{cases}\mathrm{I} & \text{for } i=1\\ \mathrm{T}_{i-1}-\mathbf{w}_{\pi(i-1)}\mathbf{w}_{\pi(i-1)}^H & \text{for } i=2, \ldots, q\end{cases}$$

and $i=1, \cdots, q$ with $q=rank(\mathbf{H}(S))$. $\mathbf{T}_i$ is the orthogonal projector matrix onto $Sp(\mathbf{h}_{\pi(j<i)})$ the subspace spanned by all previously encoded users for which $\mathbf{h}_{\pi(j<i)}\mathbf{w}_{\pi(i)}=0$. Some authors (e.g., [25], [27]) use the following expression as an objective function over the channel matrix $\mathbf{H}(S)$ for user selection and ZFDP beamforming:

$$f(\mathbf{H}(\mathrm{S}))=\sum_{i=1}^{|\mathrm{S}|}\|\mathbf{h}_i\mathbf{T}_i\|^2 \tag{18}$$

Observe that user selection and sum rate maximization based on metric (18) implicitly depend on one particular selected encoding order π out of $|S|!$ different valid permutations. Since different encoding order yield different values of (18), in [27] it was proposed a method to perform the successive encoding optimizing of the order π. Such an optimum order is attained by an iterative algorithm that evaluates (8) each iteration for every successive encoded user. An alternative suboptimal approach can be employed as in [19] where π is defined by the descending order of the effective channel gains of the users in S.

5.2. Approximation of the NSP

The objective function of problem (9) can be further relaxed by using a lower bound of $\mathrm{tr}(\dot{\mathbf{H}}(S))$ in order to avoid the computation of the inverse matrix $\dot{\mathbf{H}}(S)$. Considering the definition of trace we have that

$$\mathrm{tr}(\dot{\mathbf{H}}(\mathrm{S}))=\sum_i\lambda_i^{-1}(\bar{\mathbf{H}}(\mathrm{S})) \tag{19}$$

and using the arithmetic-geometric mean inequality over $\lambda_i(\dot{\mathbf{H}}(S))$ it holds that [28]:

$$|S|\prod_i \lambda_i^{1/|S|}(\dot{\mathbf{H}}(S)) \le \mathrm{tr}(\dot{\mathbf{H}}(S)) \tag{20}$$

Since $|S|$ is constant and independent of the selected channels for all L_{N_t} possible user sets, the lower bound on the objective function of (9) can be simplified as:

$$\prod_i \lambda_i(\dot{\mathbf{H}}(S)) = \det(\bar{\mathbf{H}}(S))^{-1} \tag{21}$$

A suboptimal but less computational demanding way to find a set of users that solves problem (4) is given by the set that solves the following combinatorial problem:

$$S_\xi = \arg \max_{S\subset\Omega:|S|=N_t} \det(\bar{\mathbf{H}}(S)) \tag{22}$$

where the optimized objective function only requires to compute the determinant of a matrix product. Observe that the lower bound in (21) is closely related with (14) and the performance degradation for the former metric arises because the terms $\det\left(\tilde{\mathbf{H}}_i\tilde{\mathbf{H}}_i^H\right)\ \forall\, i\in S$ are neglected.

In [26] it was presented a greedy algorithm where the metric for user selection is based on an approximation of (16) and the correlation coefficients are used instead of the partial correlation coefficients. Such relaxation neglects the channel gain degradation due to the terms $\pi(1)\pi(2)\cdots\pi(k-1)$. Given channel matrix $\mathbf{H}(S)$, the metric that approximates (19) is defined as follows [26]:

$$\Lambda(\mathbf{H}(S)) = \sum_i \left(\left\|\mathbf{h}_i\right\|^2 \prod_{j\neq i} \sin^2\theta_{\mathbf{h}_i\mathbf{h}_j} \right) \forall i,j\in S \tag{23}$$

Using this metric a suboptimal solution to problem (4) is given by the set of users that solve the following combinatorial problem:

$$S_\Lambda = \arg \max_{S\subset\Omega:|S|=N_t} \Lambda(\mathbf{H}(S)) \tag{24}$$

5.3. ε-orthogonality

Several user selection algorithms (e.g., [25], [26], [29]) attempt to create groups of quasi-orthogonal users based on the information provided by the coefficient of correlation (13). A set of channels $\mathbf{h}_i\ \forall\, i\in S$ is called ε-orthogonal if $\cos\theta_{\mathbf{h}_i\mathbf{h}_j} <,\ \forall\, i,\ j\in S$ [29]. Some works addressed problem (4) by scheduling the set of user with minimum ε-orthogonality measured either over the normal channels [30] or over the eigenvectors computed by SVD [29]. If the orthogonality among channels in $\mathbf{H}(S)$ is the only metric considered to define S (regardless of the channel gains), problem (4) can be sub-optimally solved by the set S that minimizes the -orthogonality among all L_{N_t} possible sets formally described as:

$$\mathrm{S}_{\delta} = \arg \min_{\mathrm{S} \subset \Omega : |\mathrm{S}| = N_t} \left(\max_{i,j \in \mathrm{S}} \cos \theta_{\mathbf{h}_i \mathbf{h}_j} \right) \tag{25}$$

Some works in the literature define S as the set with minimum sum of correlation coefficients $\sum_i \sum_{j \neq i} \cos \theta_{\mathbf{h}_i \mathbf{h}_j}$, $\forall i, j \in S$ or as the set with minimum average correlation coefficient [25], [29]. Observe that these objective functions are based on pairwise metrics and they can be negatively biased by few terms with large values neglecting the remaining coefficients with relatively small values. An alternative utility function that can identify such large terms is the geometric mean over all correlation coefficients since it would assign the same priority to each of term. In MU-MIMO systems the user grouping problem based on (25) for scheduling time slots, sub-carriers, or both, can be modeled as a graph coloring problem [13], [31] or graph clique problem [32] and complexity reduction is the main objective of the proposed grouping algorithms.

5.4. Orthogonality defect

The orthogonality defect derived from Hadamard's inequality [28] measures how close a basis is to orthogonal. Given the matrix $\mathbf{H}(S)$ this metric is given by:

$$\delta(\mathbf{H}(\mathrm{S})) = \frac{\prod_{i=1} \|\mathbf{h}_i\|^2}{\prod_{i=1} \lambda_i \left(\bar{\mathbf{H}}(\mathrm{S}) \right)} \tag{26}$$

and $\delta(\mathbf{H}(S))=1$ if and only if the elements of $\mathbf{H}(S)$ are pairwise orthogonal. The metric (26) reflects the orthogonality of the set $\{\mathbf{h}_i\}_{i \in S}$ and has been used to evaluate the compatibility between wireless channels in order to maximize the spatial multiplexing gain [29]. The original problem (4) can be sub-optimally solved by the set that minimizes the orthogonality defect which is formally described as:

$$\mathrm{S}_{\delta} = \arg \min_{\mathrm{S} \subset \Omega : |\mathrm{S}| = N_t} \delta(\mathbf{H}(\mathrm{S})) \tag{27}$$

Observe that problem (27) uses a weighted version of the utility function of problem (22) where the weight is defined by the inverse of $\prod_{i=1} \|\mathbf{h}_i\|^2$. The orthogonality defect can be seen as a scaled version of the lower bound of metric (19).

5.5. Condition number

The ZF-based beamforming methods are in general power inefficient since the spatial direction of $\mathbf{w}_i$ is not matched to $\mathbf{h}_i$. Inverting a ill-conditioned channel matrix $\mathbf{H}(S)$ yields a significant power penalty and a strong SINR degradation at the receivers. In numerical analysis a metric to measure the invertibility of a matrix is given by the condition number. In MIMO system this metric is used to measure how the eigenvalues of a channel matrix spread out and to indicate multipath richness for a given channel realization. The less spread of the eigenvalues, the larger the achievable capacity in the high SNR regime. For the matrix $\mathbf{H}(S)$ the condition number is defined as [33]:

$$\kappa(\mathbf{H}(\mathsf{S})) = \frac{|\lambda_{\max}(\bar{\mathbf{H}}(\mathsf{S}))|}{|\lambda_{\min}(\bar{\mathbf{H}}(\mathsf{S}))|} \tag{28}$$

and $\kappa(\mathbf{H}(S))$ is an indication of the multiplexing gain of a MIMO system. Another definition of the condition number is given by the product $\|\mathbf{A}\|\|\mathbf{A}^{-1}\|$ for a given non-singular square matrix $\mathbf{A}$ [20] and (28) generalizes that metric for any matrix $\mathbf{H}(S) \in \mathbb{C}^{|S|\times N_t}$ where $|S| \leq N_t$. If the condition number is small, the matrix $\mathbf{H}(S)$ is said to be well-conditioned which implies that as $\kappa(\mathbf{H}(S)) \to 1$ the total achievable sum rate in the MISO systems under ZF-based beamforming can achieve a large portion of the sum rate of the inter-user interference free scenario. Problem (4) can be sub-optimally solved by a set of users with the minimum condition number and such set is formally described as:

$$\mathsf{S}_\kappa = \arg \min_{\mathsf{S}\subset\Omega:|\mathsf{S}|=N_t} \kappa(\mathbf{H}(\mathsf{S})) \tag{29}$$

Another important metric to estimate matrix condition is given by the Demmel condition number. For such metric several applications in MIMO systems have been proposed in recent years, e.g., link adaptation, coding, and beamforming [34]. The Demmel condition number can be seen as the ratio between the total energy of the channels of $\mathbf{H}(S)$ over the magnitude of the smallest eigenvalue of $\bar{\mathbf{H}}(S)$ in the current channel realization and is given by the following expression [34]:

$$\kappa_D(\mathbf{H}(\mathsf{S})) = \frac{\mathrm{tr}(\bar{\mathbf{H}}(\mathsf{S}))}{\lambda_{\min}(\bar{\mathbf{H}}(\mathsf{S}))} \tag{30}$$

where $\mathrm{tr}(\bar{\mathbf{H}}(S)) = \|\mathbf{H}(S)\|_F^2$, i.e., the Frobenius norm is related with the overall energy of the channel. By using (30) the set of users that sub-optimally solves (4) is given by the solution of the following combinatorial problem:

$$S_{\kappa_D} = \arg \min_{S \subset \Omega : |S| = N_t} \kappa_D(\mathbf{H}(S)) \tag{31}$$

6. User selection algorithms

For MU-MIMO BC systems with ZF-based beamforming the optimum solution to the sum rate maximization problem implies the optimization over the power allocation and user selection as well as the correct dimension and orientation of the signal subspace used by each selected user [9]. Since this optimization is performed over several dimensions (time, space, signal space, power, users) the optimum solution for the MU-MIMO scenario has not been found. Even when the signal space spans over one dimension, i.e. single-antenna users, the optimum solution to the sum rate maximization problem (4) in MU-MISO BC systems is given by an exhaustive search whose computational complexity increases approximately with $O(K^{N_t})$. In order to find a fair trade-off between complexity and performance, a large number of works have addressed the sum rate maximization problem for ZFBF (e.g., [9], [21], [35] and ZFDP (e.g., [18], [19]) by designing low-complexity algorithms that find a sub-optimal yet acceptable solution to (4). The goal of such algorithms is to construct the solution set S of quasi-orthogonal users implementing different iterative approaches. Some works in the literature proposed greedy opportunistic algorithms that exploit the instantaneous channel information. In a greedy selection each new selected user finds a locally maximum for a given global objective function so that the final set of users attempt to converge to a close-to-optimal solution. Another approach to solve (4) is to use channel metrics and reformulate the original problem as an integer program. In this section we present the principles and characteristics of different user selection algorithms proposed in the literature. Our aim is to introduce generic structures of the user selection process and to illustrate under which conditions they can be used to maximize the sum rate and to reduce computational complexity for ZF-based beamforming.

6.1. Metric-based selection

The objective of the metric-based user selection is to find a set S of users with spatially quasi-orthogonal channels such that the profit from beamforming is maximized. User selection algorithms can be design to relax the original optimization problem (4) by optimizing a particular channel metric. This kind of optimization may face two main problems: 1) the large search space L for the optimum solution, and 2) how to discriminate between metrics of sets with different cardinality without computing the objective function. Both problems are partially solved by imposing a constraint in the optimizations problem, i.e., $|S| = N_t \ \forall S \subseteq \Omega$ in problem (4) or by operating in the high SNR regime. This yields a search space of size L_{N_t} and metrics that can be easily compared since they are applied over user sets of the same size. In MU-MISO BC systems with ZF-based precoding and optimal power allocation, authors in [5] showed that there exist a critical SNR that depends on $\mathbf{H}(S)$ for which $\forall P \leq P_0$ the maximum sum rate is achieved by a subset S with cardinality $|S| < rank(\mathbf{H}(S))$. For an operation point

$\forall P > P_0$ the system achieves full multiplexing diversity, i.e., $|S| = rank(\mathbf{H}(S))$. The metric-based user selection algorithms solve (4) in two phases. In the first phase S is selected in order to optimize a given metric and in the second phase the sum rate is evaluated for the previously defined set. This means that the user selection and the resource allocation problems are decoupled in this approach. The cardinality of S is fixed in the first phase and it may be modified during the second phase when the sum rate is evaluated. If power allocation is performed based on water-filling, this might yield a zero power allocation (no data transmission) for some users due to the instantaneous value of P_0.

In the literature of user selection several algorithms solve problem (4) optimizing the NSP by iteratively computing the sum of the effective channel gains each iteration. The optimum set that maximizes the NSP is given by the solution of (9) which is a combinatorial problem that requires the evaluation of L_{N_t} matrix product and inversion operations. Therefore, the set S is found in a greedy fashion by selecting the user that locally maximizes the total sum of effective channel gains which roughly requires $L_K = \sum_{k=1}^{N_t - 1} (K - k)$ evaluations of the metric $f(\bullet)$. Given $S(n-1)$ in the nth iteration, the optimum new user $i \in \Omega$ achieves the largest effective channel gain when its channel is projected onto the orthogonal subspace spanned by the previously selected users $\mathcal{V} = Sp(\mathbf{H}(S(n-1)))$. In [10], [21] the authors evaluate iteratively at the nth iteration the intersection of the null spaces of the previously selected users $S(n-1)$ with each new candidate user by means of GSO. Using this technique, it is not necessary to extract the basis of the subspace spanned by the channel matrix (SVD) or to compute the orthogonal projector matrix $\mathbf{Q}_{\mathcal{V}_i}$ at each iteration [19]. Other user selection algorithms based on the NSP further reduce the required computational complexity by approximating any of the its alternative forms described in Section 5.1.1. The algorithm proposed in [26] approximates (16) by using only the correlation coefficient between the users in $S(n)$. The authors in [36] select users based on metric (17) where each user computes its own projector matrix and the algorithm approximates the NSP. The optimization of an approximation of the NSP yields a suboptimal set of users and performance degradation regarding to (8) yet some gains in terms of complexity are attained. The general structure of this kind of algorithms is presented in Alg. 1 and $f(\bullet)$ is given by any metric defined in Section 5. Notice that the first selected user is the one with the strongest channel which not necessarily belongs to the optimum set. However, this criterion simplifies the initialization of S and does not yield a significant performance degradation, specially in the large K regime [21].

Another relevant feature of this generic structure is given by the adjustment of Ω in each iteration. This is relevant in the large K regime where the number of operations required to find the next user is roughly K which can be computationally demanding. In order to reduce the total number of metric evaluations, an optimization over Ω can be performed. For instance, by performing an optimization similar to (25) per iteration for a given ε-orthogonality target whose optimal value depends on K and N_t [37]. The principle behind this optimization is that the candidate users must satisfy the hyperslab condition $\Omega(\mathrm{n}) = \{\mathbf{h}_i \ \forall\ i \in \Omega(\mathrm{n}-1) : \cos\theta_{\mathbf{h}_i \mathbf{g}} \leq \varepsilon\}$ for a given vector $\mathbf{g}$ [21]. The metric-based user selection does not guarantee full multiplexing gain or sum rate maximization and authors in [10], [25] have proposed a second optimization

http://dx.doi.org/10.5772/58937

once that S have been found. This is referred in Alg. 1 as *Removal Optimization* and its objective is twofold: to discard inactive users that do not receive data and to maximize the achievable sum rate by optimizing the beamforming weights and powers of the active users.

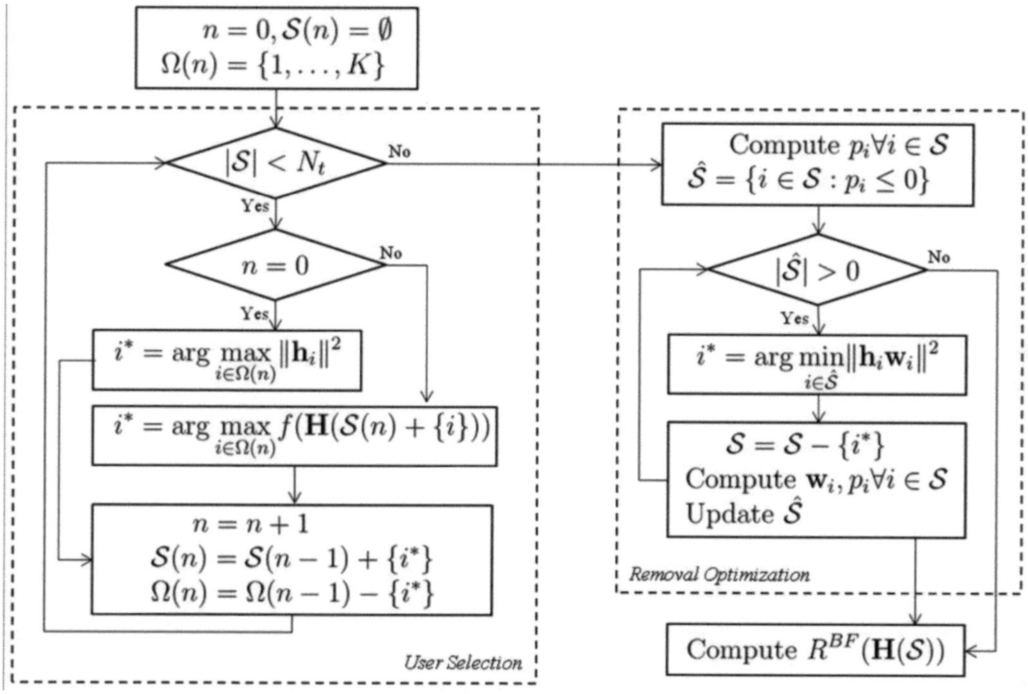

Alg 1. Generic structure of the Metric-Based User Selection

6.2. Utility-based selection

This class of algorithms perform a joint user selection and resource allocation optimization by evaluating the global utility function (3) each iteration (e.g., [10], [11], [18], [38], [39]). The greedy algorithms are highly effective for sum rate maximization and well suited for ZF-based beamforming. However, they require the evaluation of the power allocation and the achievable sum rate for each unselected user in every iteration of the algorithm which involves matrix operations (e.g., SVD or GSO). The algorithm in [18] was extended in [10] by computing an inverse matrix operation based on the LQ decomposition avoiding the explicit calculation of the Moore-Penrose pseudo-inverse, and reducing computational complexity. The general structure of the utility-based user selection is summarized in Alg. 2. Notice that as in Alg. 1, the set S is initialized with the user with maximum channel magnitude which may not be part of the optimum set. This greedy selection does not attempt to maximize the multiplexing diversity since it will stop the search as soon as the aggregation of a new local optimal user i^* decreases the overall sum rate. In the low SNR regime and for low values of K, such stop criterion highly improves the rate performance since power is allocated to strong users and there exist more degrees of freedom (DoF) at the BS to mitigate inter-users interference. This approach may face two drawbacks: 1) the achievable fairness at low SNR and low K regimes since less users are scheduled; 2) the complexity in evaluating at most L_K times the achievable sum rate. The fairness issue can be tackled by changing the global utility function, e.g., maximizing the weighted sum rate in (3) using the proportional fairness approach [21], [25], [39]. The complexity problem can be alleviated by exploiting the properties of water-filling in order to reduce the computational load of some matrix operations. Further complexity gains can be obtained if the set of candidate users Ω is optimized each iteration, reducing the total number of sum rate evaluations. This refinement proposed in [10], [18] is called *Search Space Pruning* and is based on the fact that if water-filling allocates p_i=0 to the ith user in $\Omega(n)$ at the nth iteration, such user cannot achieved a nonzero power allocation in future iterations.

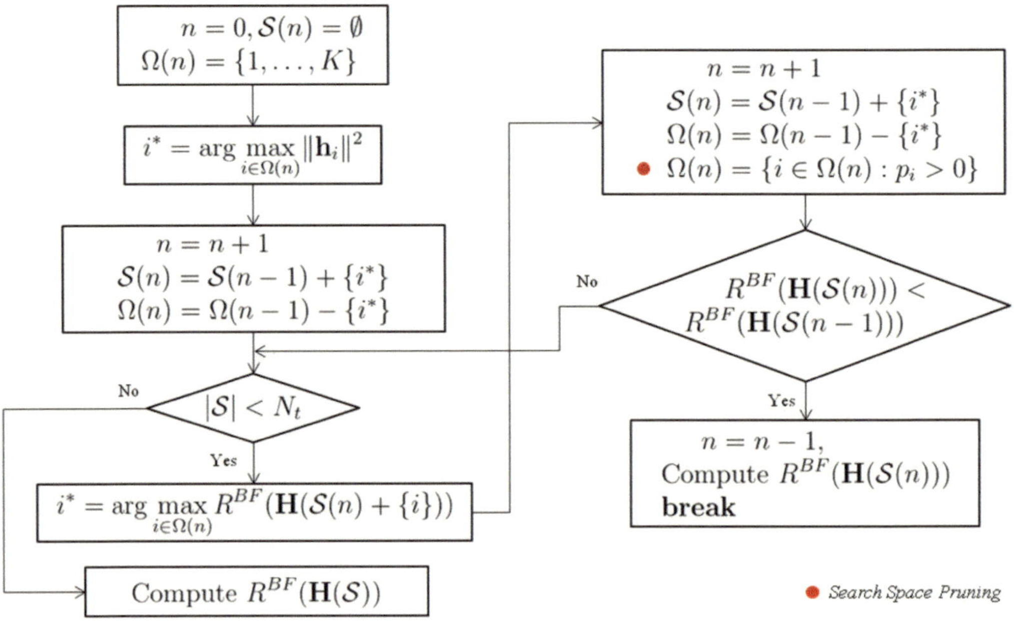

Alg. 2. Generic structure of the Utility-Based User Selection

6.3. User selection via integer optimization

The sum rate maximization formulation in (4) is a combinatorial problem subject to mixed constraints. Some works in the literature reformulate the original problem similarly to the metric-based approaches, performing user selection (S) and resource allocation ($\mathbf{w}_i$ and p_i $\forall$ $i \in S$) as independent processes [25], [26]. The NSP in Section 5.1 can be used to reformulate the utility-based selection presented in Alg. 1 as an integer program with linear objective function and linear constraint [26]. Let us consider that for each user $i \in \Omega$ there exists a function f_i that computes the NSP assuming pairwise orthogonality with user

$j \neq i, \ \forall j \in \Omega$. By computing metric (23) over all users in Ω the intersection of the projection onto the null spaces for the ith user is given by:

$$f_i = \|\mathbf{h}_i\|^2 \prod_{j \neq i} \sin^2 \theta_{\mathbf{h}_i \mathbf{h}_j} \tag{32}$$

where $\theta_{\mathbf{h}_i \mathbf{h}_j}$ can be calculated from (13). By applying a change of variables over (32) we have that $\tilde{f}_i = a_i + \sum_{j \neq i} b_{ij}$, where $a_i = 2\log(\|\mathbf{h}_i\|)$ and $b_{ij} = 2\log(\sin \theta_{\mathbf{h}_i \mathbf{h}_j})$. In this way, the component related to the channel gains and the one related to the spatial compatibility are decoupled and they can be used to model a bi-criterion problem [25] or to render the NSP into an integer problem [26]. The user selection based on the NSP seeks the maximum sum of the effective channel gains, which can be approximated by $\sum_{i \in S} \tilde{f}_i$ subject to $|S| = N_t$. In order to introduce such constraint, we define the following binary variables:

$$\phi_i = \begin{cases} 1 & \text{if user } i \text{ is selected} \\ 0 & \text{otherwise} \end{cases} \qquad \varphi_{ij} = \begin{cases} 1 & \text{if users } i \text{ and } j \text{ are selected} \\ 0 & \text{otherwise} \end{cases}$$

and the user selection modeled as an integer program based on the channel norms and correlation coefficients (13) is given by [26]:

$$\begin{aligned} \text{max.} \quad & \sum_i a_i \phi_i + 2\sum_i \sum_{j=i+1} b_{ij} \varphi_{ij} \\ \text{s.t.} \quad & \begin{cases} \sum_i \phi_i = N_t & \varphi_{ij} \leq \phi_i, \forall i,j & \varphi_{ij} \leq \phi_j, \forall i,j \\ \phi_i + \phi_j \leq 1 + \varphi_{ij}, \forall i,j & \phi_i \in \{0,1\} & \varphi_{ij} \in \{0,1\} \end{cases} \end{aligned} \tag{33}$$

Notice that the binary programming problem (33) is a reformulation of (24) and in contrast to the metric-based selection in Section 6.1, the order in which the users are selected has no impact on the orthogonality of the elements of $\mathbf{H}(S)$. The solution to the user selection problem is given by the binary variables ϕ_i so that $S = \{i \in \Omega : \phi_i = 1\}$ and power allocation based on water-filling is performed over the set of selected users according to the employed precoding scheme. Since the objective function is convex and the constraints are given by affine functions, this problem can be solved by the pseudo-dual simplex method [7] for integer programs or by using standard optimization packages like MATLAB.

7. Numerical examples

In this section we compare the average sum rate achieved when optimal and suboptimal user selection is performed over different metrics of $\mathbf{H}(S)$. The simulations consider perfect CSIT, fading channels are generated following a complex Gaussian distribution with unit variance and the average sum rate is given in (bps/Hz). Since we evaluate system performance via Shannon capacity expressions in (6) and (7), the results are independent of the specific implementation on the coding and modulation schemes, which provides us a general design insight.

The curves displayed in Fig. 3 are obtained by optimally solving the combinatorial problems introduced in Section 5 whose optimum user sets are employed to evaluate (4) with ZFBF. Notice that these results are upper bounds of the average sum rate for each metric, which implies that any greedy user selection algorithm can achieve at most the same performance for its optimized metric. The sum rate achieved by S_ω in (9) has a negligible performance gap regarding to the optimum set $S^\star$ for low values of K, and such gap rapidly vanishes as $K \to \infty$ with a cost of matrix product and inversion for each possible set $S \subseteq \Omega$. Due to the properties of water-filling the sum rate is maximized when the terms b_i have larger and uniform values. For low values of K there exists a performance gap between $S^\star$ and S_ω because the probability that the terms b_i have large uniform values is small. As K grows this probability increases and the performance gap vanishes. The average sum rate achieved by S_ξ in (22) computes a lower bound of the NSP metric (8) with a performance gap not larger than 1% regarding to $S^\star$ for $K=6$ and such gap vanishes as K grows. This result suggest that (21) is a good candidate metric for user selection since it achieves a good trade-off between complexity and performance. The fact that $\delta(\mathbf{H}(S))$ cannot identify with accuracy the set that maximizes the sum rate is because this metric only reflects the degradation of the product of the eigenvalues of $\overline{\mathbf{H}}(S)$ with respect to the channel gains. Consider that for a given channel instance the optimum solution of (27) yields $\delta(\mathbf{H}(S_\delta))=1$ which may indicate that $\mathbf{H}(S_\delta)$ forms an orthogonal basis yet this results does not provide information regarding the channel magnitudes. Observe that the optimum channel matrix $\mathbf{H}(S^\star)$ does not achieve perfect ε-orthogonality or pairwise orthogonality among its components. The sum rate attained with the solution of problems (27), (29), and (31) degrades as K grows. This is due to the fact that for large multiuser diversity the probability of finding a set of users that minimizes the dispersion of the eigenvalues of $\overline{\mathbf{H}}(S)$ increases but this does not imply that the maximum sum of effective channel gains or transmitted power is achieved. Metric (23) improves its performance when $K \to \infty$ since the probability that $\mathbf{H}(S_\Lambda)$ achieves both large sum of effective channel gains and pairwise uncorrelated channels is a function of the multiuser diversity. The performance gap between S_Λ regarding $S^\star$ ranges from 12% for $K=6$ to less than 1% for $K=100$. The advantage of metric (23) is that only the computation of correlation coefficients in (13) is required avoiding matrix operations. The performance gap between the solutions of (24) and (9) depends on the spatial resources at the transmitter since the probability that two independent channels $\mathbf{h}_i$ and $\mathbf{h}_j$ are correlated is a decreasing function of N_t. For the ε-orthogonality optimization in (25), the set S may be not unique since two different sets may containing the same maximum correlation coefficient but their achievable sum rates may be quite different. Due to the fact that channel magnitudes are neglected, problem (4) is solved without fully exploiting multiuser diversity.

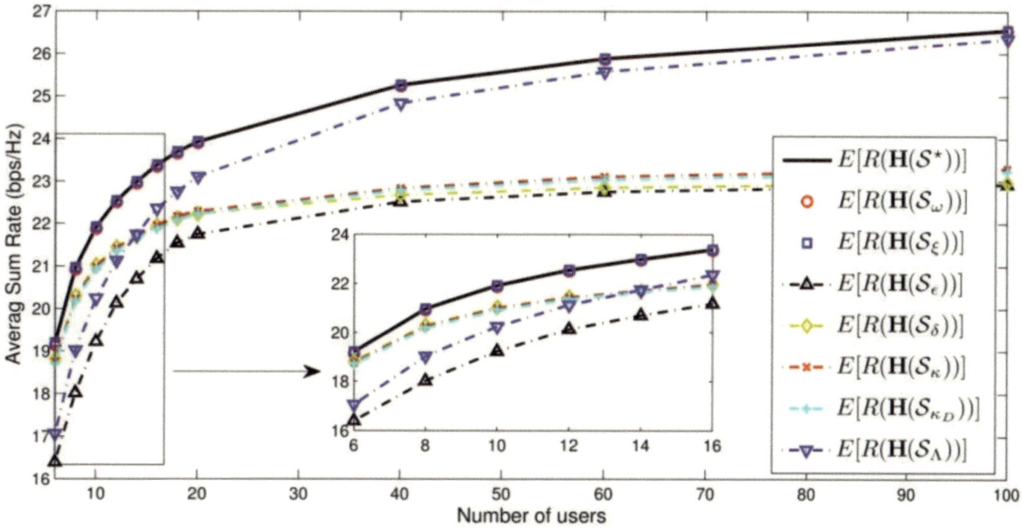

Figure 3. Average sum rate versus the number of users K with SNR=18(dB) and N_t=4.

In Fig. 4 and Fig. 5 we compare several user selection algorithms that find sub-optimal solution to (4), namely the metric-based selection using the NSP metric proposed in [21] using GSO, [19], [35] using SVD, and the NSP approximation based on (23) proposed in [26]. The optimal solution of (4) is found by exhaustive search subject to $|S| = N_t$, i.e., maximum multiplexing gain. In order to highlight the contribution of multiuser diversity we compare performance with respect to two simplistic user selection approaches, one based on the maximum channel gain (MCG) criterion (selecting the N_t users with higher channels norms), and a second approach performing round robin user scheduling (RRS) policy. We compare the performance of the metric-based selection in Section 6.1 with two greedy utility-based (sum rate) algorithms in Section 6.2 [18] and [40], and with the integer linear program (ILP) selection described in Section 6.3.

In Fig. 4 we compare the average sum rate as a function of the number of users for different user selection strategies for both precoding techniques. For the case of ZFBF Fig. 4 (a) and $K<10$, the solution set of the utility-based selection in [18] is achieved with a cardinality less than N_t compared to the optimum selection. This suggest that when $K \approx N_t$ is more efficient to select the users in a greedy fashion as described in Section 6.2, than to perform the selection based on channel metrics. When the number of competing users is large $K \gg N_t$, the performance gap between the three user selection approaches described in Section 6 decreases and all techniques achieve maximum multiplexing gain. For the large K regime, the metric-based selection achieves a performance close to the utility-based selection with less computational effort. Moreover, the performance of the NSP approximation in [26] converges to one of [19], [21], [35] with more gains in terms of computational complexity. In the case of ZFDP Fig. 4 (b) all approaches in Section 6 achieve full multiplexing diversity regardless the regime of K. This is due to the fact that ZFDP can efficiently use all spatial resources at the transmitter for the high SNR regime. The utility and metric based selection techniques achieve the same performance over all range of K. The NSP approximation metric in [26] achieves more than 95% of the optimum capacity and the performance gap reduces as the number of users increases. Notice that the approach in Section 6.3 optimally solves problem (24) and its performance is an upper bound for the user selection in [26]. The performance of all approaches presented in Section 6 is sub-optimal, yet they represent different and acceptable trade-offs between sum rate performance, computational complexity, and multiplexing gain for different values of K and the SNR regime.

In Fig. 5 the average sum rate is a function of the SNR for ZFBF and $K=10$. The results show that for $P<P_0\approx 10(\mathrm{dB})$ the utility-based selection is more efficient than the metric-based selection in the low SNR regime. In the high SNR regime the performance gap between the optimum selection and the approaches in Section 6 is less than 10%. Observe that even the NSP approximation metric in [26] can efficiently exploit multiuser diversity and the approaches based exclusively on channel magnitudes (MCG) and random selection incur in a large performance degradation. Comparing the performance of the utility-based versus the metric-based user selection is in general not fair since the former will optimize directly its utility function while the latter attempts to reduce computational complexity by optimizing a simpler utility function. The performance gap between the two approaches is large for low SRN and K regimes. The reason is that the former directly optimizes the set of selected users while the latter indirectly estimates if a given set can maximize the sum rate. In the high SNR regime, the gap between both approaches diminishes and results in [18] show that the achievable sum rate in both approaches is proportional to the number spatial resources N_t at the transmitter.

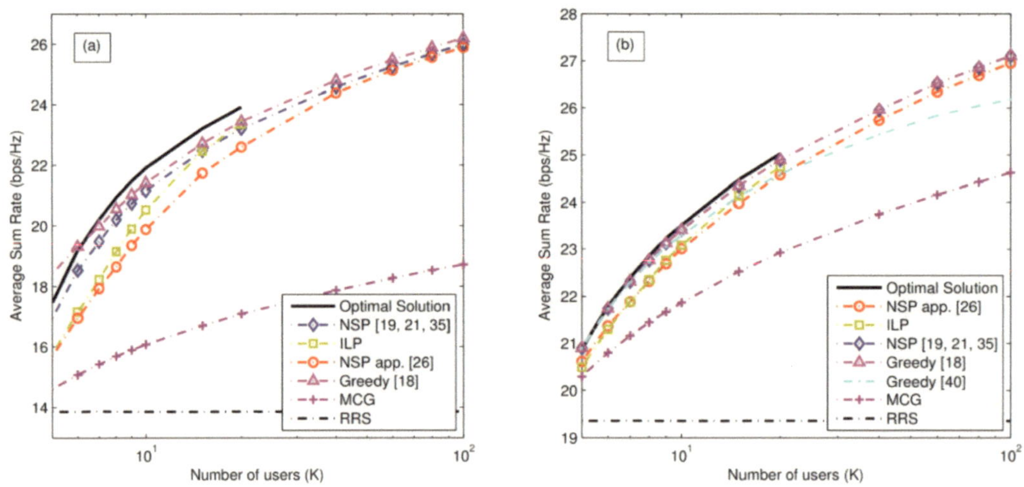

Figure 4. Average sum rate versus the number of users (K) with SNR $=18$ (dB) and $N_t=4$ for (a) ZFBF and (b) ZFDP.

http://dx.doi.org/10.5772/58937

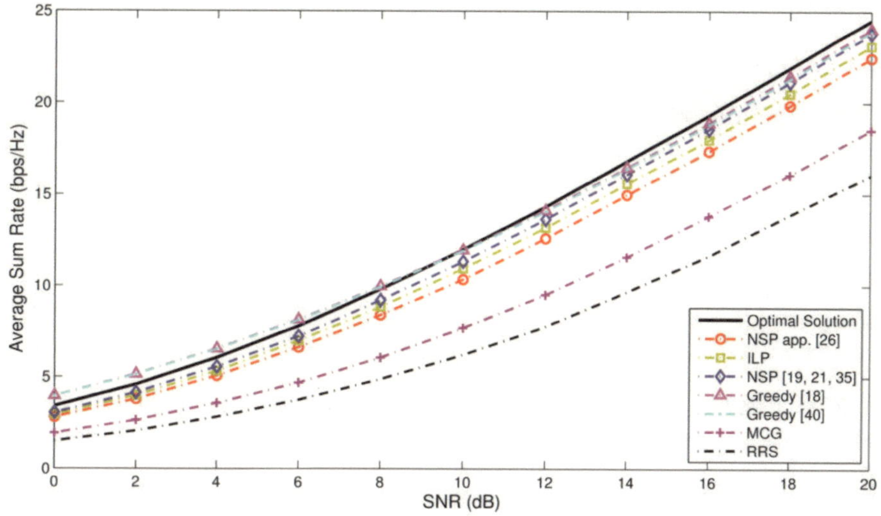

Figure 5. Average sum rate versus the SNR for the with $K=10$, $N_t=4$, and ZFBF.

8. Summary

In this chapter, we have presented several metrics and techniques for user selection in multiuser MISO scenarios. This topic has been object of extensive research over the last years for different wireless scenarios. Several metrics that evaluate spatial compatibility, matrix invertibility, pairwise orthogonality, and eigenvalue dispersion were presented and compared for the sum rate maximization with user selection problem. Results show that the null space projection and its approximations are the metrics that most efficiently identify the set of users that maximizes the sum rate for both precoders ZFBF and ZFDP. Since the optimal user selection for sum rate maximization is a complex combinatorial problem different sub-optimal selection strategies were presented. The metric-based and utility-based techniques compute the user selection in a greedy fashion finding close-to-optimal solution. The metric-based selection was reformulated as an integer program selection which provides an upper bound of the performance of a metric based on the null space projection. Numerical results show that depending on the multiuser diversity and SNR regime one strategy of user selection can be preferred over the others reaching a fair trade-off between performance and complexity.

Acknowledgements

This work was supported by the Portuguese Fundação para a Ciência e Tecnologia (FCT) ADIN (PTDC/EEI-TEL/2990/2012).

Author details

E. Castañeda, A. Silva and A. Gameiro

University of Aveiro/Instituto de Telecomunicações (IT), Portugal

References

[1] D. Tse and P. Viswanath, *Fundamentals of Wireless Communication*. Cambridge University Press, 2005.

[2] L. Zheng and D. Tse, "Diversity and multiplexing: A fundamental tradeoff in multiple-antenna channels," *IEEE Trans. Inf. Theory*, vol. 1, no. 8, pp. 1–25, 2003.

[3] M. Sharif and B. Hassibi, "A Comparison of Time-Sharing, DPC, and Beamforming for MIMO Broadcast Channels With Many Users," *IEEE Trans. Commun.*, vol. 55, no. 1, pp. 11–15, Jan. 2007.

[4] M. H. M. Costa, "Writing on dirty paper," *IEEE Trans. Inf. Theory*, vol. 29, no. 3, pp. 439–441, 1983.

[5] G. Caire and S. Shamai Shitz, "On the achievable throughput of a multiantenna Gaussian broadcast channel," *IEEE Trans. Inf. Theory*, 2003.

[6] L. C. Godara, *Handbook of Antennas in Wireless Communications*. CRC Press, 2002.

[7] S. Rao, *Engineering Optimization: Theory and Practice*. John Wiley & Sons, Inc., 2009.

[8] J. Jiang, R. Buehrer, and W. Tranter, "Greedy scheduling performance for a zero-forcing dirty-paper coded system," *IEEE Trans. Commun.*, vol. 54, no. 5, pp. 789–793, 2006.

[9] P. W. C. Chan and R. S. Cheng, "Capacity Maximization for Zero-Forcing MIMO-OFDMA Downlink Systems with Multiuser Diversity," *IEEE Trans. Wirel. Commun.*, vol. 6, no. 5, pp. 1880–1889, 2007.

[10] J. Wang, D. J. Love, and M. D. Zoltowski, "User Selection With Zero-Forcing Beamforming Achieves the Asymptotically Optimal Sum Rate," *IEEE Trans. Signal Process.*, vol. 56, no. 8, pp. 3713–3726, Aug. 2008.

[11] B. C. Lim, W. A. Krzymień, and C. Schlegel, "Efficient sum rate maximization and resource allocation in block-diagonalized space-division multiplexing," *IEEE Trans. Veh. Technol.*, vol. 58, no. 1, pp. 478–484, 2009.

[12] L.-N. Tran and E.-K. Hong, "Multiuser diversity for successive zero-forcing dirty paper coding: greedy scheduling algorithms and asymptotic performance analysis," *IEEE Trans. Signal Process.*, vol. 58, no. 6, pp. 3411–3416, 2010.

http://dx.doi.org/10.5772/58937

[13] F. Shad, T. D. Todd, V. Kezys, and J. Litva, "Dynamic slot allocation (DSA) in indoor SDMA/TDMA using a smart antenna basestation," *IEEE/ACM Trans. Netw.*, vol. 9, no. 1, pp. 69–81, 2001.

[14] E. Björnson and E. A. Jorswieck, *Optimal Resource Allocation in Coordinated Multi-Cell Systems*, vol. 9, no. 2–3. Foundations and Trends® in Communications and Information Theory, 2013, pp. 113–381.

[15] S. Christensen, R. Agarwal, E. Carvalho, and J. M. Cioffi, "Weighted sum-rate maximization using weighted MMSE for MIMO-BC beamforming design," *IEEE Trans. Wirel. Commun.*, vol. 7, no. 12, pp. 4792–4799, Dec. 2008.

[16] Y. Liu, Y. Dai, and Z.-Q. (Tom) Luo, "Coordinated beamforming for MISO interference channel: Complexity analysis and efficient algorithms," *IEEE Trans. Signal Process.*, vol. 298, no. 0704, 2011.

[17] H. Yanai, K. Takeuchi, and Y. Takane, *Projection Matrices, Generalized Inverse Matrices, and Singular Value Decomposition*. Springer, 2011.

[18] G. Dimic and N. Sidiropoulos, "On downlink beamforming with greedy user selection: performance analysis and a simple new algorithm," *IEEE Trans. Signal Process.*, vol. 53, no. 10, pp. 3857–3868, 2005.

[19] Z. Tu and R. Blum, "Multiuser diversity for a dirty paper approach," *IEEE Commun. Lett.*, pp. 1–3, 2003.

[20] J. Gentle, *Matrix algebra: theory, computations, and applications in statistics*. Springer, 2007.

[21] T. Yoo and A. Goldsmith, "On the optimality of multiantenna broadcast scheduling using zero-forcing beamforming," *IEEE J. Sel. Areas Commun.*, vol. 24, no. 3, pp. 528–541, 2006.

[22] R. Chen, R. W. Heath, and J. G. Andrews, "Transmit selection diversity for unitary precoded multiuser spatial multiplexing systems with linear receivers," *IEEE Trans. Signal Process.*, vol. 55, no. 3, pp. 1159–1171, 2007.

[23] A. Razi, D. Ryan, I. B. Collings, and J. Yuan, "Sum rates, rate allocation, and user scheduling for multi-user MIMO vector perturbation precoding," *IEEE Trans. Wirel. Commun.*, vol. 9, no. 1, pp. 356–365, 2010.

[24] I. Halperin, "The product of projection operators," *Acta Sci. Math.(Szeged)*, 1962.

[25] T. F. Maciel and A. Klein, "On the Performance, Complexity, and Fairness of Suboptimal Resource Allocation for Multiuser MIMO–OFDMA Systems," *IEEE Trans. Veh. Technol.*, vol. 59, pp. 406–419, 2010.

[26] E. D. Castañeda, A. Silva, R. Samano-Robles, and A. Gameiro, "Low-Complexity User Selection for Rate Maximization in MIMO Broadcast Channels with Downlink Beamforming.," *Sci. World J.*, vol. 2014, p. 865905, Jan. 2014.

[27] P. Tejera, W. Utschick, G. Bauch, and J. A. Nossek, "Subchannel allocation in multiuser multiple-input–multiple-output systems," *IEEE Trans. Inf. Theory*, pp. 1–34, 2006.

[28] T. Cover and J. Thomas, *Elements of Information Theory*, 2nd ed. John Wiley & Sons, Inc., 2006.

[29] A. Bayesteh and A. K. Khandani, "On the user selection for MIMO broadcast channels," *IEEE Trans. Inf. Theory*, pp. 2325–2329, 2008.

[30] Y. J. Zhang and K. Letaief, "An Efficient Resource-Allocation Scheme for Spatial Multiuser Access in MIMO/OFDM Systems," *IEEE Trans. Commun.*, vol. 53, no. 1, pp. 107–116, Jan. 2005.

[31] E. Driouch and W. Ajib, "Efficient scheduling algorithms for multiantenna CDMA systems," *IEEE Trans. Veh. Technol.*, vol. 61, no. 2, pp. 521–532, 2012.

[32] T. Yoo and A. Goldsmith, "Sum-rate optimal multi-antenna downlink beamforming strategy based on clique search," *IEEE GLOBECOM*, p. 5 pp., 2005.

[33] C. Oestges and B. Clerckx, *MIMO wireless communications: from real-world propagation to space-time code design*. 2007.

[34] C. Zhong, M. McKay, T. Ratnarajah, and K.-K. Wong, "Distribution of the Demmel condition number of Wishart matrices," *IEEE Trans. Wirel. Commun.*, vol. 59, no. 5, pp. 1309–1320, 2011.

[35] L.-C. Wang and C. Yeh, "Scheduling for multiuser MIMO broadcast systems: transmit or receive beamforming?," *IEEE Trans. Wirel. Commun.*, vol. 9, no. 9, pp. 2779–2791, 2010.

[36] M. Fuchs, G. Del Galdo, and M. Haardt, "Low-Complexity Space–Time–Frequency Scheduling for MIMO Systems With SDMA," *IEEE Trans. Veh. Technol.*, vol. 56, no. 5, pp. 2775–2784, Sep. 2007.

[37] L. Sun and M. McKay, "Eigen-based transceivers for the MIMO broadcast channel with semi-orthogonal user selection," *IEEE Trans. Signal Process.*, vol. 58, no. 10, pp. 5246–5261, 2010.

[38] Z. Shen, C. Runhua, J. G. Andrews, R. W. Heath, and B. Evans, "Low complexity user selection algorithms for multiuser MIMO systems with block diagonalization," *IEEE Trans. Signal Process.*, vol. 54, no. 9, pp. 3658–3663, 2006.

[39] L.-N. Tran, M. Bengtsson, and B. Ottersten, "Iterative Precoder Design and User Scheduling for Block-Diagonalized Systems," *IEEE Trans. Signal Process.*, vol. 60, no. 7, pp. 3726–3739, Jul. 2012.

[40] S. Karachontzitis and D. Toumpakaris, "Efficient and low-complexity user selection for the multiuser MISO downlink," *2009 IEEE 20th Int. Symp. Pers. Indoor Mob. Radio Commun.*, pp. 3094–3098, Sep. 2009.

Chapter 3

Interference Alignment — Practical Challenges and Test-bed Implementation

Nima N. Moghadam, Hamed Farhadi,
Per Zetterberg, Majid Nasiri Khormuji and
Mikael Skoglund

Additional information is available at the end of the chapter

http://dx.doi.org/10.5772/59200

1. Introduction

Data traffic over wireless communication networks has experienced a tremendous growth in the last decade, and it is predicted to exponentially increase in the next decades [1]. Enabling future wireless networks to fulfill this expectation is a challenging task both due to the scarcity of radio resources (e.g. spectrum and energy), and also the inherent characteristics of the wireless transmission medium. Wireless transmission is in general subject to two phenomena: *fading* and *interference*. The former is a consequence of reflectors scattered in the environment surrounding a transmitter and a receiver such that the receiver observes a superposition of multiple copies of the transmitted signal. The superposition of the signals can be either constructive or destructive depending on the phase shift and attenuation of the received signals from different paths. The randomness of fading may degrade communication quality. Several effective techniques have been developed to overcome the adverse effects of random fading. For instance, multiple antennas transmission techniques are proposed to realize spatial diversity and to improve the performance of wireless systems in the presence of fading [2]. In addition, in multi-user networks because of the broadcast nature of wireless transmission medium, each user's communication is interfered by other users. Inter-user interference can severely degrade the communication quality and makes communication of different users interrelated; thus, finding the optimum interference management strategy becomes a challenging problem.

Conventional interference management strategies including time-division multiple access (TDMA) and frequency-division multiple-access (FDMA) avoid the inter-user interference by allocating orthogonal resources in time and frequency to different users, respectively. Interference is consequently avoided at the cost of low spectral efficiency. Thus, it was believed that the performance of wireless networks is limited by interference in general. However, the elegant *interference alignment* concept [3], [4] reveals that with proper transmission signalling design, different interference signals can in fact be aligned together, such that more radio resources can be assigned to the desired transmission. For instance, in the case of a multi-user interference network with more than two source–destination pairs, the interference signals at each destination can be aligned such that maximally half of the signal space can be left to its desired signal [4]. Therefore, each user may achieve half of the interference-free transmission rate no matter how many interferers exist [4]. Although interference alignment can achieve a larger data rate compared to orthogonal transmission strategies, several challenges should be addressed to enable the deployment of this technique in future wireless networks [1], [5]. For instance, to perform interference alignment, normally, global channel state information (CSI) is required to be perfectly known at all terminals. Clearly, acquiring such channel knowledge is a challenging problem in practice and proper channel training and channel state feedback techniques need to be deployed. In addition, since the channels are time-varying proper adaptive transmission is needed.

To investigate whether the outstanding performance of signal processing algorithms inspired by interference alignment can be preserved in real environment, practical verifications is needed. Wireless test-beds (e.g. the ones based on USRP or WARP) can be used as a platform for the experimental verification of the novel interference management algorithms.

This chapter review recent advances in practical aspects of interference alignment. It also presents recent test-bed implementations of signal processing algorithms for the realization of interference alignment. In Section 2 we give a brief overview on the interference alignment concept. Section 3 presents the structure of a canonical transmitter and receiver to realize interference alignment, and discuss channel training and channel state feedback for these systems. A brief review on test-bed implementations of interference alignment solutions is presented in Section 4. Section 5, introduces hardware and software setup of the test-bed used in this chapter for implementation of interference alignment. The test-bed implementation of iterative transceiver design and power control algorithm is presented in Section 6. We discuss the test-bed implementation of compressed feedback scheme for interference alignment scheme in Section 7. Finally, Section 8 concludes the chapter.

2. *K*-user (*K*>2) interference networks

Consider a K-user (K>2) interference network. Several techniques have been devised for interference management in multi-user interference networks. Three major approaches to deal with interference are illustrated in Fig. 1. In Fig. 1 (a) all sources simultaneously transmit their signals in the same frequency band. Each source applies the conventional single-user coding technique. At each destination, the desired signal is superimposed by interference signals. The destination performs decoding by treating the interference signals as noise. This strategy is

http://dx.doi.org/10.5772/59200

reasonable for cases that the receiver only knows the transmitted codebooks of the intended source. In the low signal-to-noise ratio (SNR) region, the level of interference can be controlled by proper power control techniques. However, in the high-SNR regime, inter-user interference is dominant. Therefore, the power control alone is not sufficient to manage the interference and this transmission strategy may not lead to a desirable performance.

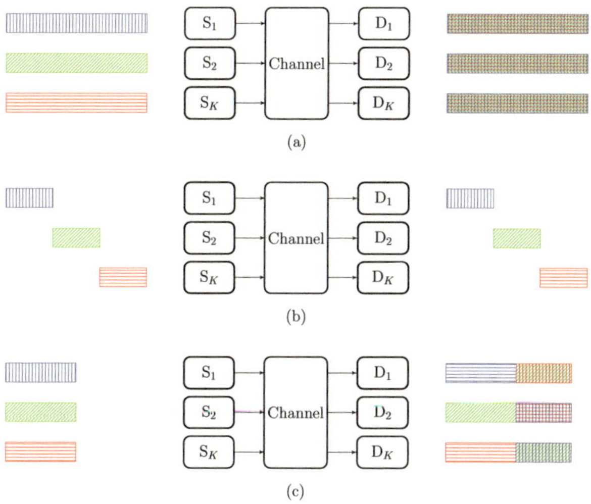

Figure 1. Transmission schemes in three-user interference networks: (a) non-orthogonal transmission and decdoing by treating interference as noise, (b) orthogonal transmission, and (c) interference alignment.

The conventional approach to avoid interference at destinations is to orthogonalize the transmissions of different users. Each source–destination pair has access to only a portion of the available channel, as shown in Fig. 1 (b). Although signal reception at each destination does not directly suffer from inter-user interference, this scheme is not spectrally efficient due to the fact the resource; i.e, time or bandwidth, are divided among the source–destination pairs. From Fig. 1 (b), we see that the interference signals span a large dimension of the received signal space at each destination to ensure orthogonal reception. However, if at each destination the dimension of the subspace occupied by only the interference signals can be reduced, a larger interference-free subspace would be left for desired transmission. This can be realized using a new technique called *interference alignment* [3].

Specifically, interference alignment for interference networks refers to "*a construction of signals in such a manner that they cast overlapping shadows at the receivers where they constitute interference while they remain distinguishable at the intended receivers where they are desired*" [4]. In general, two conditions should be satisfied to perform interference alignment technique. First, interference signals should be aligned at the same subspace, termed *interference subspace*. Next, we need to check whether the subspace left for the desired signal, called *desired subspace*, is independent from the interference subspace. The conditions are essentially required for the realization of the interference alignment techniques. An illustrative representation of this concept is shown in Fig. 1 (c).

Interference alignment can be realized in different domains such as space (across multiple antennas [3], [4]), time (exploiting propagation delays [6], [7] or coding across time-varying channels [4], [8]), frequency (coding across different carriers in frequency-selective channels [9]), and code (aligning interference in signal levels [10]). Combinations of domains can also be used e.g. space and frequency, [11]. In the following, we briefly introduce *Degrees of Freedom (DoF)* which is a performance measure for wireless networks at high-SNR regime.

2.1. Degrees of freedom region

Consider the K-user interference network in Fig. 1. Source S_k ($k \in \{1, 2, \ldots, K\}$) intends to send an independent message $w_k \in W_k$ to its destination, where W_k denotes the corresponding message set. The message $|w_k|$ is encoded to a codeword of length N. Thus, the corresponding code rate is $R_k = \frac{\log|w_k|}{n}$ where $|w_k|$ denotes the cardinality of $|w_k|$. The rate tuple (R_1, R_2,..., R_K) is said to be achievable if a sequence of codebooks exists, such that the probability that each destination decodes its message in error can be arbitrarily small, by choosing long enough codewords. The capacity region of the network is the closure of the set of all achievable rates. In Gaussian interference networks where the noise is additive white Gaussian, the capacity region depends on the transmission powers of sources, the noise powers and channel gains. Since the exact capacity region is difficult to find, as a starting point one can use the DoF region to characterize/approximate the capacity/achievable rate region in the high-SNR region (where interference is the dominant phenomenon that degrades system performance). The DoF region is defined as follows

$$\mathsf{D} = \left\{ (d_1, \ldots, d_K) \in \mathbb{R}^+ \mid \exists (R_1, \ldots, R_K) \in \mathcal{C}(P), d_k = \lim_{P \to \infty} \frac{R_k}{\log P}, 1 \le k \le K \right\}, \tag{1}$$

where *C(P)* denotes the capacity region, and P is the transmission power of each source. The DoF can be seen as the pre-log factor of the achievable rate and the DoF region describes how the capacity region expands as transmission power increases.

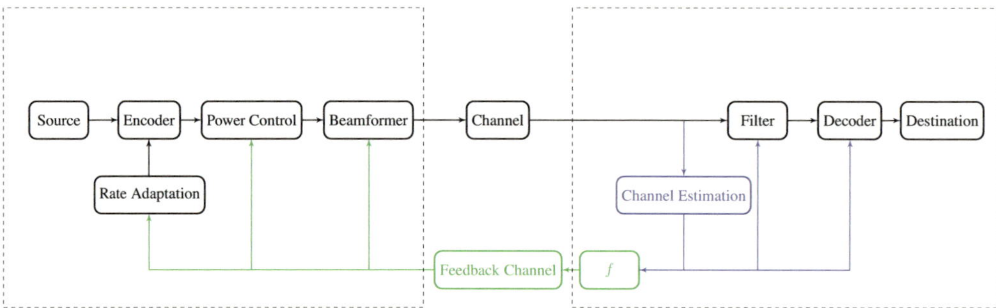

Figure 2. Transmitter and Receiver Structure.

http://dx.doi.org/10.5772/59200

3. Practical challenges of interference alignment

The structure of a canonical transmitter and receiver for the implementation of interference alignment is shown in Fig. 2. At the transmitter side, there is an *encoder* which encodes the messages to the corresponding codewords suitable for transmission over the channel. The transmission can be enhanced by the adaptation of the transmitted signal according to the received channel state infomation feedback. Specifically, in a class of communication systems that transmission powers are fixed and a maximum throughput is desired, the encoder may adapt transmission rate according to the estimate of the mutual information of the channel (computed by the *rate adaptation* module). On the other hand, in another class of systems which desire fixed-rate transmission, *power control* module should adjust transmitted power according to the channel state feedback to maintain mutual information of the channel larger than a certain level. Each transmitter has a *beamformer* which compute the proper signal for transmission over the channel according to the interference alignment concept.

At the receiver side, *channel estimation* module computes the estimation of incoming channel gains. These channel estimations can be used for recovering the transmitted message and computing the channel state information feedback signal. The *filter* module exploits estimated channel gains to recover the desired signal from interference signals according to the interference alignment concept. The *decoder* module decodes the message using an estimate of the incoming channel gain. The feedback encoder module denoted by 'f' in Fig. 2 computes the feedback signal according to the estimated channel gains. Also, there is a *synchronizer* module at the receiver to synchronize the receiver and the transmitter. In the following, we will explain these parts in more detail.

3.1. Channel training for interference alignment

In practice, destinations can acquire CSI through a pilot-based channel training scheme. For example, consider block fading channel model in which channel gains are constant within one fading block and change to independent values in the subsequent blocks. The length of each block coincides with the coherence time of channel denoted as T.

Terminals first need to learn the channels within each fading block, and next use the estimated channels to perform their transmission. A pilot-assisted interference alignment scheme is proposed in [12] which perform these tasks. According to this scheme, transmission within each fading block is conducted in two phases: *pilot transmission phase* and *data transmission phase*. These two phases have the duration of $T_\tau=\alpha T$ and $T_d=(1-\alpha)T$, respectively, where $\alpha \in [K/T, 1]$ called *channel sharing factor* is a design parameter. A larger α leads to more accurate channel estimation at the expense of less time left for data transmission. The transmission power of the pilot symbols in general can be different from that of the data symbols. Let P, P_d, and P_τ denote the average transmission power, the average power of data symbols, and the average power of pilot symbols, respectively. Source S_l ($l \in \{1,..., K\}$) sends T_τ known pilot symbols with power P_τ. Then, the destination D_k ($k \in \{1,..., K\}$) applies a minimum mean square error (MMSE) estimator to obtain an estimate of the channel gain. As shown in Fig. 2 this channel estimate is used at the receiver filter and the decoder to recover the desired message [13].

A more accurate estimation of the channel can be obtain by allocating more transmission power for training symbols which implies that a lower power is left for data transmission. The achievable rate region by taking into account this noisy CSI is computed in [12]. According to Proposition 2 in [12], the optimum power allocation which maximizes the achievable rate region is

$$P_{d,\text{opt}} = \beta_{\text{opt}} P, P_{\tau,\text{opt}} = K\left(\left(1-(1-\alpha)\beta_{\text{opt}}\right)/\alpha\right)P, \tag{2}$$

where

$$\beta_{\text{opt}} = \frac{1}{1-\alpha}\left(1+\sqrt{\frac{1+KP/(1-\alpha)}{1+PT/N_0}}\right)^{-1}. \tag{3}$$

If $P \gg 1$, then the

$$\beta_{\text{opt}} \approx \frac{1/(1-\alpha)}{1+\sqrt{KN_0/T(1-\alpha)}}. \tag{4}$$

This result recommends that in large networks more power should be allocated to the channel training instead of the data transmission. The intuition behind this result is that in large networks the performance loss due to imperfect interference alignment as a consequence of imperfect CSI becomes more important. Thus, it is recommended to allocate more power to pilot symbols instead of data symbols to acquire CSI more accurately.

3.2. Channel state information feedback

As we have discussed in the previous section, destinations can acquire CSI through a pilot-based channel training scheme. The destinations then can send the estimated CSI to the sources via channel state feedbacks. They can transmit either un-quantized CSI (analog feedback) or quantized CSI (digital feedback) via feedback channels. In the following, we briefly review some key results for different type of feedback schemes.

1) Analog Feedback: The destinations can obtain an estimate of the incoming channels according to the scheme mentioned in Section 3.1. Then, they may transmit the analog value of the estimated channels over the feedback channel. Let the function f in Fig. 2 to be defined as $f(x)=x$, and assume error-free feedback channels. According to Theorem 3 in [12], in a single-input single-output (SISO) time-varying K-user network with coherence time T, the achievable sum DoF is

$$d_{\Sigma} = \min\{K(1-K/T)/2, T/8\}. \tag{5}$$

This result is achieved when the number of the users selected to be active is

$$K_{\text{opt}} = \min\{K, T/2\}. \tag{6}$$

This recommends that, in large networks ($K > T/2$) when no CSI is a priori available at terminals, first select a subset of the users and next perform channel training and interference alignment within the set of active users.

2) Digital Feedback: Digital channel state feedback strategies in which each destination quantize the incoming channels and sends the index of the quantized channel over feedback channels has been studied in several works (e.g. [14-16]). It is shown that the same DoF as when perfect CSI is available can be obtained, provided that the destinations a priori know channels and the feedback signals' rate is proportional to log P, where P is the transmission power of each source. That is,

$$N_f \sim \log P, \tag{7}$$

where N_f is the number of feedback bits. It should be noticed that in practice where destinations do not know incoming channels a priori, part of the radio resources need to be allocated for channel training and a smaller DoF will be achievable (cf. [12]). This part is discussed in more detail in section 7.1.

3.3. Iterative interference alignment

In this part we present an iterative algorithm referred to as *leakage minimization algorithm* and an extension for it called *Max-SINR* which both are proposed in [17]. Consider a three-user multiple-input multiple-output (MIMO) interference network where each terminal is equipped with M antennas. For presentation simplicity here we assume M to be even. The result when M is odd is provided in [4]. It has been shown in [4] that the achievable DoF of each source–destination pair is $M/2$. To achieve this DoF, the transmitter-side beamformers and the receiver-side filters should be designed. In the following, we will briefly present the algorithm proposed in [17] to compute the beamformers and filters. We assume each terminal can acquire only local channel side information, i.e. knowledge about the channels which are directly connected to it throughout training of the channel. Destination D_k can obtain $\mathbf{H}_{kl}$ and source S_k can obtain $\mathbf{H}_{lk}$, $l \in \{1, 2, 3\}$, where $\mathbf{H}_{lk}$ is the forward channel from S_k to D_l and $\mathbf{H}^{R_{lk}}$ is the reverse channel from D_l to S_k. The iterative computation of the beamformers and the filters occur in the training phase. After the convergence of the computed filters and beamformers to the interference alignment solutions, the data transmission starts. Let $\mathbf{V}_k$ denote an $M \times M/2$ transmitter-side beamforming matrix where its columns are the orthogonal basis of the transmitted signal space of S_k. The transmitted signal of S_k is

$$\mathbf{x}_k = \mathbf{V}_k \bar{\mathbf{x}}_k, \tag{8}$$

where each element of the $M/2 \times 1$ vector $\bar{x}_k$ represents an independently encoded Gaussian codeword with power $2P_k/M$. Each codeword is beamformed with the corresponding column of $\mathbf{V}_k$.

Let $\mathbf{U}_k$ be an $M \times M/2$ receiver-side filtering matrix whose columns are the orthogonal basis of the desired signal subspace at D_k. The filter output of D_k is

$$\bar{\mathbf{y}}_k = \mathbf{U}_k^* \mathbf{y}_k, \tag{9}$$

where $\mathbf{y}_k$ is the received vector. The transmitter-side beamforming matrices and the receiver-side filtering matrices should satisfy the following interference alignment conditions:

$$\begin{aligned} \mathbf{U}_k^* \mathbf{H}_{kj} \mathbf{V}_j &= 0, \forall j \neq k : j,k \in \{1,2,3\}, \\ \text{rank}\left(\mathbf{U}_k^* \mathbf{H}_{kk} \mathbf{V}_k\right) &= \frac{M}{2}, \forall k : k \in \{1,2,3\}. \end{aligned} \tag{10}$$

If global CSI is available, the beamforming and filtering matrices can be designed such that these conditions are satisfied. However, with the lack of global CSI if we choose the beamformers and the filters randomly, with high probability only the second condition in (10) will be satisfied. Consequently, some interference remains at the destinations. The total power of interference at D_k is

$$IF_k = \text{Tr}\left[\mathbf{U}_k^* \mathbf{Q}_k \mathbf{U}_k\right], \tag{11}$$

where

$$\mathbf{Q}_k = \sum_{j=1, j \neq k}^{K} \frac{2P_j}{M} \mathbf{H}_{kj} \mathbf{V}_j \mathbf{V}_j^* \mathbf{H}_{kj}^* \tag{12}$$

is the covariance matrix of interference at D_k. Clearly, IF_k=0 only if the beamformers and the filters satisfy conditions in (10). However, D_k can utilize the local channel side information to minimize this received interference power by optimizing its receiver-side filter. Therefore, assuming that the beamformers are fixed, the receiver-side filter $\mathbf{U}_k$ is the solution of the following problem:

$$\min_{\mathbf{U}_k, \mathbf{U}_k \mathbf{U}_k^* = \mathbf{I}_{\frac{M}{2}}} IF_k. \tag{13}$$

The solution is given in [17]:

$$\mathbf{U}_k^d = \mathbf{v}^d[\mathbf{Q}_k], \quad d = 1, ..., \frac{M}{2}, \tag{14}$$

where $\mathbf{U}_k^d$ denotes the dth column of $\mathbf{U}_k$ and ν^d [**A**] is the eigenvector corresponding to the dth lowest eigenvalue of **A**.

To optimize the transmitter-side beamformers the reciprocity of the channels can be exploited to obtain CSI at sources. For instance, destinations can transmit training sequences over the *reverse channels* (channels from destinations to sources) which are separated from the *forward channels* (channels from sources to destinations) in time via time-division duplex (TDD). The *reciprocity* of the forward and reverse channels is assumed, i.e. $\mathbf{H}_{kl}^r = \mathbf{H}_{lk}^*(\forall l, \ k \in \{1, 2, ..., K\})$. For this purpose, the destination D_k performs beamforming and S_k perform filtering in the reverse direction with matrices $\mathbf{V}_k^d$ and $\mathbf{U}_k$, respectively. These matrices can be selected to perform interference alignment in the reverse direction. Since, the reciprocity holds, if we choose $\mathbf{U}_k^r = \mathbf{V}_k$ and $\mathbf{V}_k^r = \mathbf{U}_k$, the solutions of the interference alignment problem in the reverse direction are equivalent to those in the forward direction. In a similar way as in the forward direction, the reverse direction filters are computed as,

$$(\mathbf{U}_k^r)^d = \mathbf{v}^d[\mathbf{Q}_k^r], \quad d = 1, ..., \frac{M}{2}. \tag{15}$$

where $\mathbf{Q}^r_k$ is the covariance matrix of interference at Sk and is computed in a similar way as in (12) with the reverse direction channels and beamformers. We can set the beamforming matrices in the forward direction as $\mathbf{V}_k = \mathbf{U}_k^r$ and repeat this optimization procedure until the beamforming matrices and the receiving filters converge. The convergence of this algorithm is shown in [17]. Next, the sources beamform their data using the computed beamforming matrices and the destinations decode their desired signals by applying associated filters. An extension of this iterative algorithm is proposed in [17] which instead of minimizing leakage at each destination tries to maximize signal-to-noise-plus-interference ratio (SINR) corresponding to each srteam. This algorithm is referred to as *Max-SINR algorithm* in the literature. According to *Max-SINR algorithm*, the receiver filtering vector $\mathbf{U}_k^d$ instead of the one in (14) is selected to be a MMSE filter as follows

$$\mathbf{U}_k^d = \frac{\left(\mathbf{Q}_k^d\right)^{-1} \mathbf{H}_{kk} \mathbf{V}_k^d}{\left\| \left(\mathbf{Q}_k^d\right)^{-1} \mathbf{H}_{kk} \mathbf{V}_k^d \right\|_2}, \tag{16}$$

$$\mathbf{Q}_k^d = \sum_{j=1}^{K}\sum_{l=1}^{M/2} \frac{2P_j}{M} \mathbf{H}_{kj} \mathbf{V}_j^l \left(\mathbf{V}_j^l\right)^* \mathbf{H}_{kj}^* - \frac{2P_j}{M} \mathbf{H}_{kk} \mathbf{V}_k^d \left(\mathbf{V}_k^d\right)^* \mathbf{H}_{kk}^* + N_0 \mathbf{I}_{M/2}. \tag{17}$$

where $\mathbf{I}_{M/2}$ is M/2 × M/2 identity matrix and N_0 is the noise power. Similarly, the transmitter beamforming vectors are updated in the reverse transmissions. In the following sections, we will present test-bed implementation of algorithms which use Max-SINR algorithm for computing the transmitter and receiver filters.

4. Review on test-bed implementation of interference alignment

The idea of squeezing aligned interference signals into half of the signal space and accessing the other half of the signal space for desired transmission in an interference network is so tempting that a large body of works has been done since the introduction of interference alignment to implement this elegant approach, for instance look at [9], [18-26].

The first implementation of interference alignment is reported in [18]. A hybrid version of interference alignment combined with the successive interference cancellation (IAC) in a single carrier narrow-band MIMO wireless local area network (WLAN) is tested in this paper. Several interference alignment and interference alignment-like approaches are tested in MIMO-OFDM interference channels in [19]. This paper specifically studies the effects of poor channel conditions on the performance of interference alignment. Real-time implementation of different interference alignment scenarios are performed on different test-beds like the ones in [21-23]. [21] identifies practical issues that degrades the interference alignment performance such as channel estimation errors or collinearity between the desired signal and interference subspaces while in [23], in Vienna MIMO test-bed (VMTB), the typical delay is measured.

Implementation of interference alignment in frequency domain over measured channels is considered in [9] where the different variants of interference alignment are compared with frequency planning scenarios. The significant superiority of interference alignment performance in terms of average sum rate is reported at high SNRs in this paper.

In all previously mentioned papers, the effects of hardware impairments are ignored. In [20], the effects of such impairments on the performance of interference alignment and coordinated multi-point[1] (CoMP) is investigated on *KTH four-multi* test-bed and a measurement based model for signal-to-interference-noise-and-distortion ratio (SINDR) is suggested. Implementation of interference alignment with power control and interference alignment with compressed feedback are also studied on the same test-bed in [24] and [25], respectively. These two approaches are further explained in the rest of this chapter. Table below summarizes some of the works that have been done in the test-bed implementation of interference alignment.

1 CoMP is an approach similar to interference alignment with this main difference that in CoMP all the sources know the information to be transmitted to all destinations.

Reference	alignment dimension	implemented algorithms	number of user pairs	channels	DL/UL	antennas per node	hardware	f_c (GHz)	synchronization	Measurement environment
[18]	Space	IA in DL and IAC in UL	2 and 3	flat-fading MIMO	DL and UL	2	USRP (one per node)	–	–	indoor office
[19]	Space	variants of IA	3	MIMO-OFDM	DL	2	NI PXI-1045	2.4	high precision OCXO software correction trigger signal	indoor and out-door
[20]	Space	Max-SINR	3	MIMO-OFDM	DL	2	USRP	2.49	10 MHz CLK, NMEA, PPS	indoor office
[9]	Frequency	variants of IA	3	measured SISO-OFDM	DL	1	LTE BS	2.7	N/A	urban macrocell
[21]	Space	IA	3	MIMO-OFDM	DL	2	Lyrtech	5	40 MHz OSC hardware trigger	indoor office
[22]	Space	IA	3	MIMO-OFDM	DL	2	USRP	2.4	10 MHz CLK PPS	indoor office
[23]	Space	IA	3 (1 physical and 2 virtual Rx)	MIMO-OFDM	DL	4	VMTB	2.503	rubidium CLK, GPS	outdoor to indoor, indoor to outdoor
[24]	Space	IA with PC	3	MIMO-OFDM	DL	2	USRP	2.4	10 MHz CLK, NMEA, PPS	indoor office
[22]	Space	IA with compressed feedback	3	MIMO-OFDM	DL	2	USRP	2.4	10 MHz CLK, NMEA, PPS	indoor office

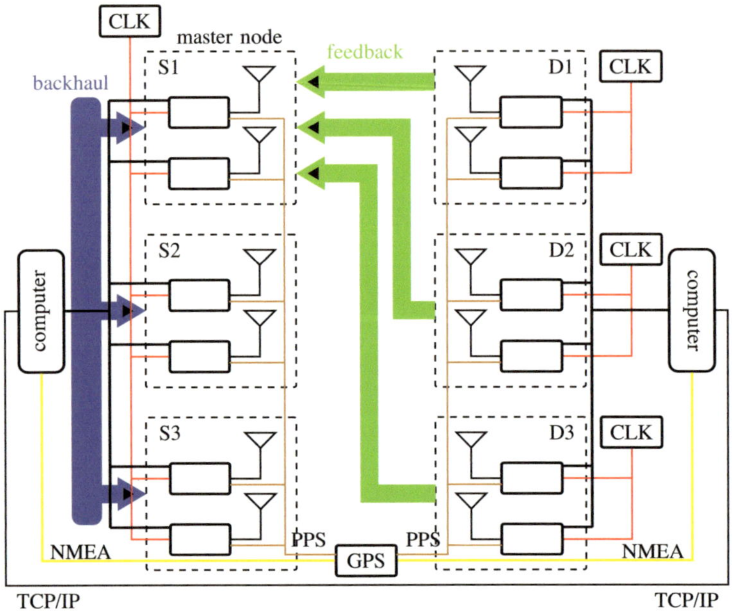

Figure 3. Network structure of four-multi test-bed.

5. KTH four-multi test-bed setup

KTH four-multi is a USRP-based wireless communication test-bed consisting of several stationary and movable multi-antenna nodes. A software framework accompanies the hardware setup of the test-bed which facilitates the rapid testing of multi-antenna schemes (see http://fourmulti.sourceforge.net/). In the following, the hardware and software structure of the test-bed is described.

5.1. Hardware setup

The current version of the test-bed consists of six nodes where three of them are fixed and take the role of transmitting sources while the other three are movable receiving destinations. All the nodes are equipped with two vertically polarized dipole antennas spaced 20 cm apart which is 1.6 times of the carrier's wavelength. Twelve Ettus Research USRP N210 (see www.ettus.com) are used to govern the twelve antennas in the network. The source USRPs are equipped with the standard Ettus XCVR2450 RF dautherboards while the destination USRPs use custom boards to achieve sufficient noise figure and dynamic range. The output signal of each source USRP is amplified by a ZRL-2400LN power amplifier. Two Linux computers control all the USRPs in the network. The network structure of KTH four-multi test-bed is illustrated in Fig. 3.

The network is designed to work at 2.49 GHz center frequency with 12 MHz bandwidth. Synchronization of the network is performed in three levels, namely time, frequency and transmit-receive synchronizations. The time and transmit-receive synchronizations are done by means of a pulse-per-second (PPS) signal (0-5 V, 1 Hz square wave) and a national marine electronics association (NMEA) signal (an ASCII protocol that provides hour-minute-second time), respectively. Both signals are generated by an EM406A GPS module and distributed through the network. The frequency synchronization is also performed by helps of 10 MHz reference clocks (CLK). All the source's local oscillators are locked to the same clock while a separate clock is provided for each of the destinations. In a real implementation the same synchronization would be achieved using common control and synchronization channels (cellular systems) or from the burst preambles (wireless local area networks). In a system with interference alignment, transmitter will in any case need some kind of back-haul to provide a common time reference and disperse scheduling decisions.

5.2. Software setup

The four-multi software framework has been developed in C++(see http://fourmulti.source-forge. net/). It runs on two Linux computers separately. One of the computers controls the three source nodes while the other one controls the three destination nodes connected to them via Ethernet connections. The sources' computer generates the transmitted frames and feeds them to the source nodes while the destinations' computer process the received frames at the destination nodes. A TCP/IP connection between the source and the destination computers provides the feedback links. Backhaul communication among the source nodes is also implemented by the help of TCP/IP connections between the source computer and the source nodes.

The framework contains a toolbox for coding and modulation (AMC and OFDM1) which was used in the implementations of the next two sections. The KTH four-multi Modulation and coding toolbox includes an LDPC channel encoder/decoder, a QAM modulator/ demodulator and an OFDM modulator/demodulator. The specification of these built-in functions is summarized in Table 1.

OFDM1		AMC	
FFT length	80	Coding rates	1/2, 5/8, 3/4
Cyclic prefix length	12	Codeword length	1520
Number of null subcarriers	42	QAM modulation orders	4, 16, 64, 256
Subcarrier spacing (KHz)	312.5		

Table 1. Modulation and coding toolbox specifications.

6. Test-bed implementation of the iterative transceiver filter design and power control

We, in this section, first present an iterative algorithm for joint transceiver filter design and power control proposed in [27] and then explain how this algorithm is implemented on KTH four-multi test-bed and finally present measurement results.

6.1. Iterative transceiver filter design and power control algorithm

In an interference network, each user can affect the received SINR at its corresponding destination through the choice of beamforming and receiving filters as well as the transmitting power. Considering single stream transmissions, the received SINR at the kth destination is computed as

$$\mathrm{SINR}_k\left(\mathbf{u}_k, \mathbf{v}_k, P_k\right) = \frac{P_k \mathbf{u}_k^* \mathbf{H}_{kk} \mathbf{v}_k \mathbf{v}_k^* \mathbf{H}_{kk}^* \mathbf{u}_k}{IF_k + N_0} \tag{18}$$

where IF_k is given by (11). Given the beamforming and receiving filters, from Equation (18) the minimum transmitting power required to maintain a fixed rate $R_{tar,k}$ (assuming that the interference can be regarded as Gaussian noise) at kth destination can be found as

$$P_k = \beta_k\left(\mathbf{u}_k, \mathbf{v}_k, \gamma_k\right) := \frac{\gamma_k\left(IF_k + N_0\right)}{\mathbf{u}_k^* \mathbf{H}_{kk} \mathbf{v}_k \mathbf{v}_k^* \mathbf{H}_{kk}^* \mathbf{u}_k}, \tag{19}$$

where $\gamma_k = 2^{R_{tar,k}} - 1$ is the minimum required SNR in an AWGN channel. The total power of the interference, IF_k which appears in the nominator is a function of transmitting powers at all the interfering transmitters. It means that increasing the power of one source causes higher level of interference at non-corresponding destinations and therefore the other sources need to transmit with higher power as well. In [28], it is shown that if the relation between the powers P_k, k=1,..., K satisfies a set of conditions then either there is an unique solution for the powers that can be found iteratively starting from any initial value or no solution exists. It is easy to prove that Equation (19) meets the conditions in [28] and therefore optimal transmitting powers can be found distributively after a sufficient number of iterations.

Inspired by Max-SINR algorithm and the aforementioned power control algorithm, an iterative transceiver design and power control algorithm was proposed in [27]. A brief version of the algorithm is presented on the next page for the sake of completeness. The algorithm is composed of three update phases in each iteration such that the receiving filters, transmission powers and beamforming filters are sequentially updated. The receiving and beamforming filters are optimized to deliver the maximum SINR at the destinations in the forward communication direction and at the sources in the reverse direction, respectively according to the concept of Max-SINR algorithm. On the other hand, in the power update phase, the powers are set to the minimum values needed for maintaining a fixed rate communication. The transmission power is upper bounded by $P_{max.}$ This algorithm assumes that accurate CSI is obtained at terminals. An extensions of the algorithm when terminals have access to only noisy CSI is proposed in [29].

Algorithm 1 Transceiver Filter Design and Power Control [27]

Initialize:$\mathbf{v}_k(0),\ P_k(0),\ k\in\{1,\ ...,\ K\},\ n=1.$

repeat

Update receiver filtering vector:

$$\mathbf{u}_k(n)=\arg\max_{u_k}\left\{\mathrm{SINR}_k\left(\mathbf{u}_k,\mathbf{v}_k(n-1),P_k(n-1)\right)\right\} \tag{20}$$

Update transmission power:

$$P_k(n)=\min\left\{\beta_k\left(\mathbf{u}_k(n),\mathbf{v}_k(n-1),\gamma_k\right),P_{\max}\right\} \tag{21}$$

Reverse the communication direction: $\begin{cases}\mathbf{u}_k^r(n-1)=\mathbf{v}_k(n)\\ \mathbf{H}_{ij}^r=\mathbf{H}^*_{ji}i,\ j\in\{1,\ ...,\ K\}\\ P_k^r(n-1)=P_F\end{cases}$

Update transmitter beamforming vector:

$$\mathbf{u}_k^r(n)=\arg\max_{\mathbf{u}_k^r}\left\{\mathrm{SINR}_k^r\left(\mathbf{u}_k^r,\mathbf{v}_k^r(n-1),P_k^r(n-1)\right)\right\} \tag{22}$$

Set beamforming vector $\mathbf{v}_k(n)=\mathbf{u}_k^r(n)$ and $n=n+1$

until$n=N$

6.2. Transmitted frame structure

The air interface of the network is designed based on OFDM modulation using KTH four-multi's modulation and coding toolbox. Coding rate of 1/2 and 16QAM modulation was chosen for transmitting fixed-rate data streams though the air interface. The transmitted frame structure is depicted in Fig. 4. In our experiment, each frame consists of 20 payload symbols and either two or three reference signals (RS) (i.e. pilot symbols). The payload symbols are *concurrently* transmitted from all the sources while each stream is beamformed with its corresponding beamforming filter instructed by Algorithm 1. Although the overhead of pilot symbols is significant in this case, we note that the number of payload symbols could be larger depending on the coherence time of the indoor channel. In the implementation in Section 7 there are in fact thousands of payload symbols with no additional pilots. Thus the overhead could be reduced to one percent for the environment in Fig. 5.

Three types of RS are employed in the network, which are referred to as *channel state information RS (CSI-RS)*, *demodulation RS (DM-RS)* and *power RS (P-RS)*. During the pilot transmission all the sub-carriers of the OFDM symbol is filled with known QAM symbols. We next explain each type of these reference signals:

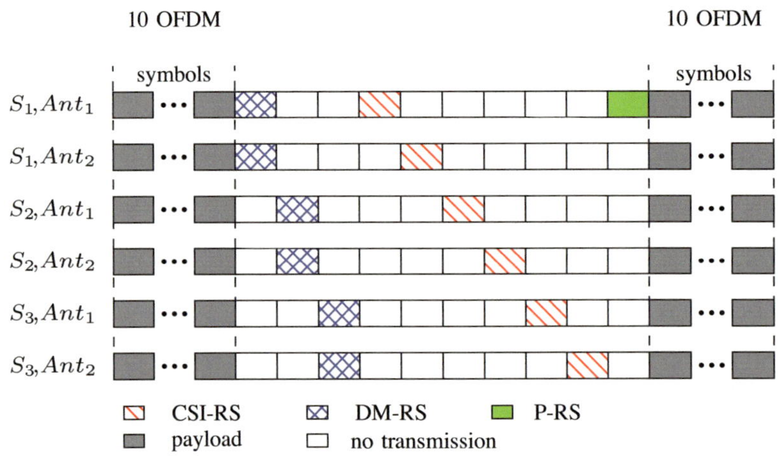

Figure 4. Transmitted frame structure in IA-PC scheme.

CSI-RS: The received noisy CSI-RS at the destinations are exploited to estimate the corresponding channel matrices to enable execution of Algorithm 1. The CSI-RS are transmitted *orthogonally*; i.e., one CSI-RS is transmitted from each transmit antenna in the network while the other antennas are silent. To enhance the quality of the channel estimations the CSI-RS are scaled such that the associated QAM symbol has the maximum transmit power P_{max}.

DM-RS: The DM-RS are used to compute the effective channel by taking into account the transmit and receive filters. Therefore they need to be stream-dedicated and be processed by the same pre-coder as the payload symbols of the corresponding stream. In this way, their power is not fixed and is set by the power control algorithm.

P-RS: Algorithm 1 is constructed to select the minimum possible transmission power to minimize the interference at the destinations. This hence reduces the power of DM-RS and may lead to a poor estimation of cross-channels, which is not favorable. To tackle this problem, P-RS is introduced where the amplitudes of the CSI-RS are scaled after the pre-coder by a scaling factor α. In each frame, the scaling factor is computed as

http://dx.doi.org/10.5772/59200

$$\alpha = \sqrt{\frac{P_{\max}}{\max_{k,j} P_{k,j}}}, \tag{23}$$

where $P_{k,j}$ is the transmit power of the jth sub-carrier of kth source. The destinations therefore need to get informed about the scaling factor. This is achieved by having node 1 (the master node) repeat its first DM-RS but now scaled with α. This enables all destinations to make robust estimate of α (it is assumed that all destinations can hear node 1). The factor α is also quantized into a discrete set of values to avoid that α introduces estimation errors.

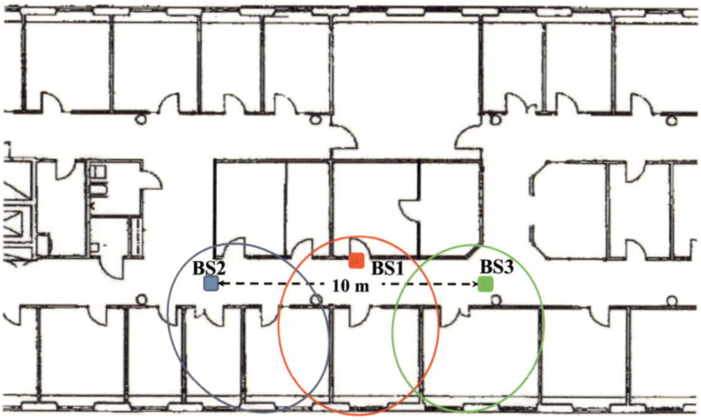

Figure 5. Measurement environment map.

6.3. Measurement results

The test-bed measurement was performed in KTH signal processing department which floor map is illustrated in Fig. 5. The measurement environment is categorized as an indoor office. In this experiment only non-line-of-sight (non-LOS) scenarios were investigated by placing source and destination nodes in the corridor and inside the nearby offices, receptively. The receive antenna gains also decreased by connecting 10 dB attenuators to them in order to avoid saturation of receive power amplifiers. The measurement was done in 100 batches. In each batch a random placement of the destination nodes in the area marked by colored circles in the figure were measured. The signals transmitted according to two different schemes were measured sequentially in each batch. In the first scheme, referred to as *noPC*, the iterative interference alignment was implemented according to [17] for benchmarking. In the second scheme, referred to as *PC*, transmission powers and beamforming filters were computed according to Algorithm 1 and MMSE receiving filters were applied at the destinations. Each scheme was run with 28 frames inter-spaced 0.15 seconds. The statistics of the first frames of both schemes were not taken into account since there is no feedback information at these frames.

High power may push the terminals' power amplifier to work in their non-linear region. Non-linearities in the transmit-receive chain degrades the performance of the system by introducing *distortion noise* into the system. Distortion noise is usually modeled as a Gaussian noise whose power increases by increasing the transmission power of the source nodes [30]. In order to make sure that the reduction of transmit power is not the only cause for the performance improvement, four different levels of transmission power were tested in noPC scheme that is each 7 frames were transmitted with a different power.

Table 2 shows the average performance of the two schemes for the 100 measurement batches. In this measurement the PC scheme's target SINR γ_k was set to 18 dB for all the users. The performances were compared in the sense of bit-error rate (BER) and transmitted power. The table shows that the PC scheme has the lowest BER, although its average transmit power is much lower than the noPC scheme with the best BER performance.

scheme	noPC				PC
Average power (dBm)	-12.9	-3.4	1.1	7.1	-1
FER	0.6856	0.1700	0.0528	0.0561	0.0071
BER	0.0815	0.0124	0.0020	0.0030	2.2 × 10–4
Average SINR (dB)	10.9	20	24.3	26.7	18.5

Table 3. Measurement results.

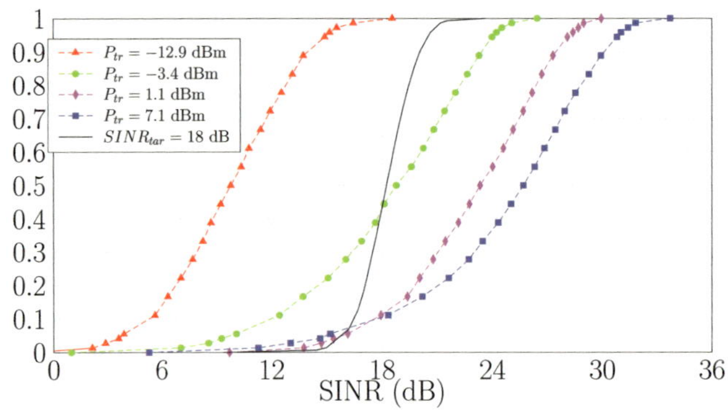

Figure 6. Empirical CDF of SINR. The solid line represents the PC scheme and the dashed lines denote the noPC scheme with different transmit powers (P_{tr}).

Empirical cumulative distribution function (CDF) of the received SINR for two schemes is plotted in Fig 6. This plot reveals the reason for the low BER of PC scheme, despite its low transmit power. The received SINRs in this scheme are concentrated around the target value while in the noPC scheme they are distributed over a wider range. Having SINR higher than the target value while the transmit rate is fixed leads to the waste of energy and on the other hand SINRs lower than the target increase the probability of error and therefore the BER.

Power saving gain at each frame is computed as the ratio of the transmit power in noPC scheme with 7.1 dBm average power and the total power transmitted in the PC scheme with 18 dB target SINR. Fig. 7 shows the empirical CDF of the power saving gain. As the empirical CDF implies, implementation of Algorithm 1 in PC scheme leads to at least 4dB gain in 90% of the measurements. Fig. 7 also shows that in 10% of the measurements gains higher than 13 dB was observed.

The benefit of power control in the PC scheme is in fact two-fold. By decreasing the transmit power, while retaining the target SINR, not only less interference is received at the destinations but also the distortion noise due to transceiver impairments decreases.

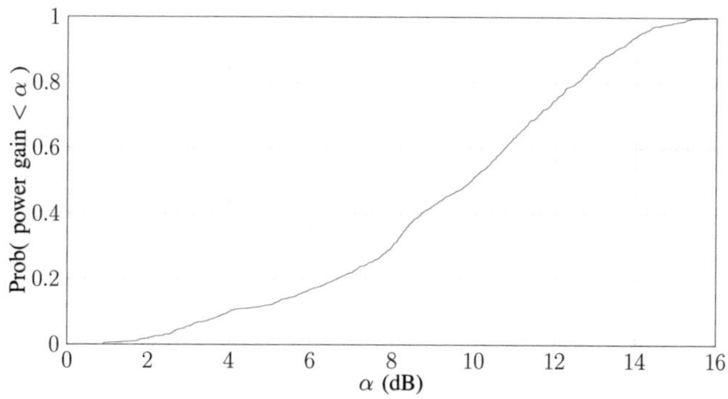

Figure 7. Emperical CDF of transmit power saving gain.

7. Test-bed implementation of the interference alignment with compressed feedback

In this section we describe the implementation of interference using limited digital feedback (see Section 3.2). Rather than using a uniform quantizer, we use a modified form of the MIMO matrix compression of the IEEE802.11ac standard. The propagation scenario and test-bed is the same as the one used in Section 6 i.e. a system with three (K=3), 2×2 MIMO links. However, the transmission power is higher than in previous sections and the system is therefore limited by interference and hardware impairments rather than thermal noise [25], [26]. A second difference from the previously presented results in Section 6, is the performance criterion. Rather than using BER we here use transmission rate as the criterion. The transmission rate is obtained by probing each link with ten different coding and modulation schemes (MCS). The rate is determined by finding the MCS with the highest rate which does not incur any frame errors.

The section is organised as follows. Section 7.1 describes the feedback compression of the CSI assuming a system with K links and M antennas in each source and destination node, while the measurement results are presented in Section 7.2.

7.1. Compression of IEEE802.11ac and adaptation to interference alignment

The feedback scheme described in the standard IEEE802.11ac resembles the feedback method for slowly time-varying single-user MIMO channels presented in [31]. In this scheme a singular value decomposition (SVD) of the MIMO channel **H** (for a certain sub-carrier) is first performed as

$$\mathbf{H} = \mathbf{W}\Lambda\mathbf{F}^*. \tag{24}$$

The destination then feeds back, in compressed form, the complex unitary matrix **F** and the real diagonal matrix **Λ**, while the complex unitary matrix **W** does not need to be fed back. To realize this we may imagine that each destination pre-multiplies its received signal with $\mathbf{W}^*$. The effective channel then becomes $\breve{\mathbf{H}} = \Lambda\mathbf{F}^*$ which is a channel which can be reconstructed by the sources. In practice, the destination will not pre-multiply its received signal with $\mathbf{W}^*$, since almost every receive technique is invariant to such a linear unitary transform.

In the case of centralised interference alignment, knowledge of the channels between all sources and destinations is required to obtain all the beamformers $\mathbf{V}_k$. We assume here that sources are connected to a common back-bone for exchange of CSI and synchronization. If we directly apply the compression scheme presented above for the single-user case so that each destination D_k, compresses the matrices $\mathbf{H}_{k,l}$ for $l \in \{1,\ldots, K\}$ it is clear that there is no single unitary matrix to be applied at the destination to transform all involved K channels into $\breve{\mathbf{H}}_{k,l} = \Lambda_{k,l}\mathbf{F}^*_{k,l}$ such that all channels can be obtained from the compressed feedback from user k.

To overcome this problem, in the system implementation we have based the feedback from destination k on the big matrix, $\mathbf{H}^{[k]}$ obtained by concatenating the channel matrices $\mathbf{H}_{k,l}$ for l=1,..., K column-wise. Thus in this case $\mathbf{H}^{[k]}$, is computed as,

$$\mathbf{H}^{[k]} = \left[\mathbf{H}_{k,1}, \ldots, \mathbf{H}_{k,K}\right]. \tag{25}$$

Thus destination D_k now feedback a compressed version of the right-hand side eigenvectors of $\mathbf{H}^{[k]}$ i.e. the real diagonal matrix $\Lambda^{[k]}$ and the complex unitary matrix $\mathbf{F}^{[k]}$ of the SVD

$$\mathbf{H}^{[k]} = \mathbf{W}^{[k]}\mathbf{L}^{[k]}\left(\mathbf{F}^{[k]}\right)^*. \tag{26}$$

The size of matrix $\Lambda^{[k]}$ is $M \times M$ while $\mathbf{F}^{[k]}$ is $KM \times M$. Just as in the single-link case we can now imagine that the destination pre-multiplies its received signal with $(\mathbf{W}^{[k]})^*$. The transmitter side is then able to obtain knowledge of the effective channel $\breve{\mathbf{H}}^{[k]} = \Lambda^{[k]}\left(\mathbf{F}^{[k]}\right)^*$. From this combined channel the constituent sub-channels can be pulled out from the corresponding columns e.g. the second channel $\breve{\mathbf{H}}_{k,2}$ corresponds to columns M+1,..., 2M of $\breve{\mathbf{H}}^{[k]}$. The reconstructed channels can then be used in place of the actual channels in any transmit beamformer and receive filter algorithm.

The IEEE 802.11ac feedback compression scheme starts by rotating the phase of the columns of $\mathbf{F}^{[k]}$ in order the last row of this matrix to have real and positive elements. These phase rotations do not need to be sent to the sources, since these rotations only amount to a phase rotation of the signals which can be undone at the destination. In the next step, $\mathbf{F}^{[k]}$, is multiplied by a diagonal matrix to have the first column become real and positive as

$$\mathbf{F}^{[k]} \leftarrow \mathbf{F}^{[k]}\ \mathbf{diag}(\exp(j\phi_{1,1}),\ldots,\exp(j\phi_{KM-1,1}),1). \tag{27}$$

These angles $\phi_{1,1},\ldots,\phi_{KM-1,1}$ are then quantised uniformly, each with b_ϕ bits. Then, real-valued Givens rotations are utilised to successively zero out the second to the KM-th element of the first column of $\mathbf{F}^{[k]}$. For the latter rotations the angles $\psi_{2,1},\ldots,\psi_{KM,1}$ are used. The angles lie between 0 and $\pi/2$ and are quantised uniformly with b_ψ bits. This procedure is repeated in a similar fashion for the remaining columns of $\mathbf{F}^{[k]}$. More details are presented in [31] and [32] as well as in the Matlab/Octave functions available at http://people.kth.se/~perz/packV/.

Since we use OFDM there is one $\mathbf{H}^{[k]}$ matrix for each user and sub-carrier. In the IEEE 802.11ac standard, the parameter N_g is defined. This parameter determines the frequency domain granularity of the feedback. If N_g=1, the feedback of $\mathbf{F}^{[k]}$ is done on every sub-carrier. If N_g=2, the feedback is only done on every other sub-carrier. In the standard, the values 1, 2 and 4 have been defined for N_g. In our measurements the values 8, 16 and 38 have also been considered, since this would significantly reduce the number of feedback bits. The angle resolution parameters b_ϕ and b_ψ are defined in the IEEE 802.11ac standard as either b_ϕ=5 and b_ψ=7 or b_ϕ=7 and b_ψ=9. In the presented results only the latter value pair is used. The total number of bits required to feed back the $\mathbf{F}^{[k]}$ matrix is given by

$$n_{\mathrm{b}} = \left((2KM-1)M - M^2\right)\left(b_\phi + b_\psi\right)/2 \tag{28}$$

The number of bits can be reduced by a further $(K-1)b_\phi$ bits. We can see this by dividing the $\mathbf{F}^{[k]}$ into sub-matrices of size $M \times M$ as

$$\left(\mathbf{F}^{[k]}\right)^* = \left[\left(\mathbf{F}_1^{[k]}\right)^*,\ldots,\left(\mathbf{F}_K^{[k]}\right)^*\right]. \tag{29}$$

Since the signals from source, l, only propagates through sub-matrix $\mathbf{F}^{[k]}_l$, we may freely multiply sub-matrices $1,\ldots,K-1$ by one phasor each. By doing so, we may set $\phi_{1,1},\ldots,\phi_{1+(K-2)M,1}$ to zero, and thus we do not need to transmit them over the feedback channels. The last sub-matrix can not be rotated since then the last row of $\mathbf{F}^{[k]}$ would no longer be real and positive.

The elements of $\Lambda[k]$ are divided by the noise standard deviation. The so obtained value can be regarded as the SNR of a corresponding stream. The reporting of these SNRs is done separately per stream, and is done in two steps. In the first step, the average SNR in dB over the whole frequency band, i.e. for all singular values of the narrowband channels for which SNR is reported, is fed back to S_k. This value is uniformly quantised with 8 bits in the range from 10 dB to 53.75 dB. In the second step, the difference in dB between the SNR of the reported sub-carrier and the average SNR is computed and is uniformly quantised with 4 bits in the range from 8 dB to 7 dB. In S_k, the SNRs of all the reported sub-carriers are first reconstructed. Then, linear interpolation (in dB) is deployed to obtain the SNR for all the sub-carriers for which $\mathbf{F}^{[k]}$ is fed back. With these two entities at hand, the channel matrices $\mathbf{H}^{[k]}$ are reconstructed and used to compute the desired pre-coders. For the sub-carriers where there is no feedback available, the pre-coding of the nearest reported sub-carrier is utilized, i.e. no matrix interpolation method is used.

A practical problem which occurred during the early experimentation was that the SNR sometimes exceeded 53.75 dB. This happened due to the high transmission power and short range. First this was handled by reporting the maximum value 53.75 dB whenever this happened. However, this resulted in making the reconstructed channel at the transmitter high-rank although the estimated channel at the receiver were in fact low-rank-which was very detrimental to the performance. To circumvent this problem, an offset is subtracted from SNR values when this condition happens.

7.2. Results

In addition to interference alignment[2], we also run the reference scheme single-user MIMO. In this scheme two spatial streams are transmitted from each source using its two transmitter antennas. Two versions of the single-user MIMO is used. In one version one source is active at a time. This version is naturally called TDMA. In another version of single-user MIMO all links are active at the same time. This scheme will work well if the inter-link interference is weak. We call this approach *full-reuse MIMO*. Finally, we also include a scheme called *full-reuse SIMO*. In this approach all sources transmit at the same time but using only a single-antenna at the sources. The difference between this scheme and interference alignment is the beamforming at the sources and the channel state feedback. All other aspects of the signal processing are identical. The performance of the system were investigated in 43 different locations for each of the three destination terminals. Half of the locations were in the corridor and half in the adjacent rooms, see Figure 5. In each location the performance was investigated with the frequency domain granularity parameter, N_g set sequentially to the values 1, 2, 4, 8, 15, 38. No person or object was moving in the environment and all schemes were run in rapid sequence thus making it possible to reliably compare the results. The results are shown in Figure 8. The dashed lines are the results obtained without the feedback compression while the solid lines are with compression (although the frequency domain granularity is still applied). The results

2 By interference alignment we are here referring to the modified form of interference alignment known as Max-SINR, see Section 3.3. However, measurements we have performed have shown that the performance difference between the original interference alignment and Max-SINR is neglible in our scenario.

http://dx.doi.org/10.5772/59200

show negligible performance loss from the quantization. However, using a too high frequency domain granularity, $N_g > 8$ or 2.5 MHz, do incur some loss. Note that the result for full-reuse MIMO is very poor showing that the interference is strong. The gain of interference alignment over the best reference scheme (TDMA MIMO) is 27%. We may also note that interference alignment provides a gain over full-reuse SIMO which proves that the transmitter beamforming is making a difference.

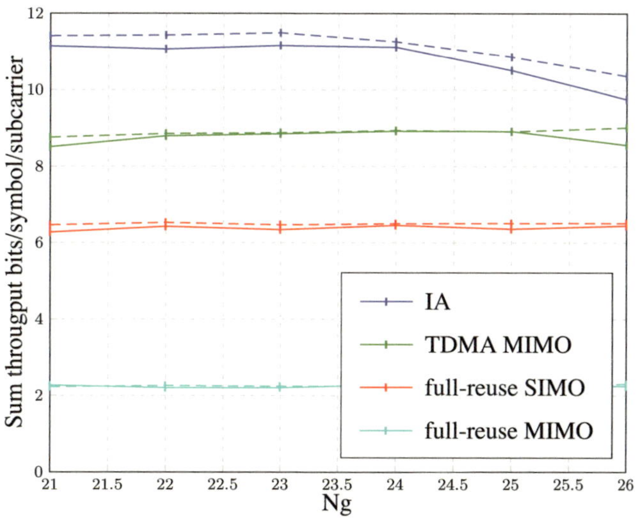

Figure 8. Measured sum throughput. The dotted line are the results without compression

When the channel is changing the performance of interference alignment will inevitably be degraded due the channel mismatch between the channel at the feedback time and the channel at the actual transmission. The amount of degradation depends on the time delay and the level of movement in the environment. In [26] using measurements we showed that the throughput of interference alignment dropped 6.4% when two passers-by were walking in the measurement environment and the feedback delay interval was 23 ms. With this delay interval the overhead of performing the feedback scheme above with N_g=8 is 2.5%, assuming that the feedback scheme of IEEE802.11ac is used.

The channel may also change due to the existence of the users at the destinations. This effect was studied by measuring the performance with and without a user located close to the destination nodes. The results were obtained using eight destination positions. The performance of interference alignment dropped 4.9% while the performance of single-user MIMO was unaffected.

These results show that interference alignment can still deliver a net performance gain of sum throughput some 15% over single-user MIMO in a WiFi scenario with three access points and three users, even when feedback overheads and mobility is taken into account.

8. Conclusion

We have reviewed the concept of interference alignment, with theoretical results regarding power and time allocation between training and payload data transmission, and previous works within experimentation on interference alignment. We further present implementation efforts addressing the combination of iterative interference alignment and power control and using compressed channel state feedback based on a modified form of the MU-MIMO feedback scheme of IEEE802.11ac. Our experimental results show that the iterative interference alignment and power control scheme is able to provide better FER/BER performance (16QAM code rate 0.5), than an implementation without power control. The results using the modified IEEE802.11ac feedback scheme, show that interference alignment can bring an improvement in throughput when considering the loss of bandwidth needed for feedback of the channel state information even for channels with realistic indoor mobility.

Acknowledgements

The research leading to these results has received funding from Swedish Foundation for Strategic Research (SSF) under RAMCOORAN project and ACCESS Linnaeus Center under graduate course wireless experimentations. The measurements were performed within the framework of the HIATUS project. The project HIATUS acknowledges the financial support of the Future and Emerging Technologies (FET) programme within the Seventh Framework Programme for Research of the European Commission under FET-Open grant number: 265578.

Author details

Nima N. Moghadam[1*], Hamed Farhadi[2], Per Zetterberg[3], Majid Nasiri Khormuji[4] and Mikael Skoglund[5]

*Address all correspondence to: nimanm@ee.kth.se

1 Department of Signal Processing, KTH, Stockholm, Sweden

2 Department of Communication Theory, KTH, Stockholm, Sweden

3 Department of Signal Processing, KTH, Stockholm, Sweden

4 Huawei Technologies Sweden AB, Stockholm, Sweden

5 Department of Communication Theory, KTH, Stockholm, Sweden

References

[1] Osseiran, A. *et al.* The foundation of the mobile and wireless communications system for 2020 and beyond challenges, enablers and technology solutions. In *IEEE Vehicular Technology Conference (VTC'13)* (2013).

[2] Tarokh, V., Seshadri, N. & Calderbank, A. R. Space-time codes for high data rate wireless communication: performance criterion and code construction. *IEEE Transactions on Information Theory* 44, 744–765 (1998).

[3] Maddah-Ali, M. A., Motahari, A. S. & Khandani, A. K. Communication over MIMO X channels: Interference alignment, decomposition, and performance analysis. *IEEE Transactions on Information Theory* 54, 3457–3470 (2008).

[4] Cadambe, V. R. & Jafar, S. A. Interference alignment and degrees of freedom of the *K*-user interference channel. *IEEE Transactions on Information Theory* 54, 3425 –3441 (2008).

[5] Ayach, O. E., Peters, S. W. & Heath, R. J. The practical challenges of interference alignment. *IEEE Transactions on Wireless Communications* 20, 35–42 (2013).

[6] Maleki, H., Jafar, S. A. & Shamai, S. Retrospective interference alignment over interference networks. *IEEE Journal of Selected Topics in Signal Processing* 6, 228–240 (2012).

[7] Maddah-Ali, M. A. & Tse, D. Completely stale transmitter channel state information is still very useful. In *48th Annual Allerton Conference on Communication, Control, and Computing (Allerton'10)* (2010).

[8] Nazer, B., Jafar, S., Gaspar, M. & Vishwanath, S. Ergodic interference alignment. In *IEEE International Symposium on Information Theory (ISIT'09)* (Seoul, Korea, 2009).

[9] Brandt, R., Asplund, H. & Bengtsson, M. Interference alignment in frequency-a measurement based performance analysis. In *Systems, Signals and Image Processing (IWSSIP), 2012 19th International Conference on*, 227–230 (2012).

[10] Motahari, A. S., Gharan, S. O., Maddah-Ali, M. A. & Khandani, A. K. Real interference alignment: Exploiting the potential of single antenna systems (2009). URL http://arxiv.org/abs/0908.2282/.

[11] Brandt, R., Zetterberg, P. & Bengtsson, M. Interference alignment over a combination of space and frequency. In *Communications Workshops (ICC), 2013 IEEE International Conference on*, 149–153 (2013).

[12] Farhadi, H., Khormuji, M. N. & Skoglund, M. Pilot-assisted ergodic interference alignment for wireless networks. In *IEEE International Conference on Acoustics, Speech and Signal Processing (ICASSP'14)* (Florence, Italy, 2014).

[13] Farhadi, H., Khormuji, M. N., Wang, C. & Skoglund, M. Ergodic interference alignment with noisy channel state information. In *IEEE International Symposium on Information Theory Proceedings (ISIT'13)* (Istanbul, Turkey, 2013).

[14] BolcSkei, H. & Thukral, I. J. Interference alignment with limited feedback. In *IEEE International Symposium on Information Theory (ISIT'09)* (Seoul, Korea, 2009).

[15] Krishnamachari, R. T. & Varanasi, M. K. Interference alignment under limited feedback for MIMO interference channels. In *IEEE International Symposium on Information Theory (ISIT'10)* (Austin, Texas, U.S.A, 2010).

[16] Kim, J. S., Moon, S. H., Lee, S. R. & Lee, I. A new channel quantization strategy for MIMO interference alignment with limited feedback. *IEEE Transactions on Wireless Communications* 11, 358–366 (2012).

[17] Gomadam, K., Cadambe, V. R. & Jafar, S. A. A distributed numerical approach to interference alignment and applications to wireless interference networks. *IEEE Transactions on Information Theory* 57, 3309 – 3322 (2011).

[18] Gollakota, S., Perli, S. D. & Katabi, D. Interference alignment and cancellation. In *In Proceedings of ACM SIGCOMM* (2009).

[19] El Ayach, O., Peters, S. & Heath, R. The feasibility of interference alignment over measured MIMO-OFDM channels. *Vehicular Technology, IEEE Transactions on* 59, 4309–4321 (2010).

[20] Zetterberg, P. & Moghadam, N. An experimental investigation of SIMO, MIMO, interference-alignment (IA) and coordinated multi-point (CoMP). In *Systems, Signals and Image Processing (IWSSIP), 2012 19th International Conference on*, 211–216 (2012).

[21] Gonzalez, O., Ramirez, D., Santamaria, I., Garcia-Naya, J. & Castedo, L. Experimental validation of interference alignment techniques using a multiuser MIMO test-bed. In *Smart Antennas (WSA), 2011 International ITG Workshop on*, 1–8 (2011).

[22] Massey, J. et al. Implementation of a real-time wireless interference alignment network. In Signals, Systems and Computers (ASILOMAR), 2012 Conference Record of the Forty Sixth Asilomar Conference on, 104–108 (2012).

[23] Mayer, M., Artner, G., Hannak, G., Lerch, M. & Guillaud, M. Measurement based evaluation of interference alignment on the vienna MIMO test-bed. In *Wireless Communication Systems (ISWCS 2013), Proceedings of the Tenth International Symposium on*, 1–5 (2013).

[24] Moghadam, N. N., Farhadi, H., Zetterberg, P. & Skoglund, M. Test-bed implementation of iterative interference alignment and power control for wireless MIMO interference networks. In *IEEE International Workshop on Signal Processing Advances in Wireless Communications (SPAWC'14)* (Toronto, Canada, 2014).

[25] Zetterberg, P. Interference alignment (IA) and coordinated multi-point (CoMP) with IEEE802.11ac feedback compression: test-bed results (2014).

[26] Zetterberg, P. Interference alignment (IA) and coordinated multi-point (CoMP) with IEEE802.11ac feedback compression: test-bed results. In *IEEE Vehicular Technology Conference* (2014).

[27] Farhadi, H., Wang, C. & Skoglund, M. Distributed interference alignment and power control for wireless MIMO interference networks. In *IEEE Wireless Communications and Networking Conference (WCNC3)* (Shanghai, China, 2013).

[28] Yates, R. A framework for uplink power control in cellular radio systems. *Selected Areas in Communications, IEEE Journal on* 13, 1341–1347 (1995).

[29] Farhadi, H., Zaidi, A., Fischione, C., Wang, C. & Skoglund, M. Distributed interference alignment and power control for wireless MIMO interference networks with noisy channel state information. In *IEEE International Black Sea Conference on Communications and Networking (BlackSeaCom'13)* (Batumi, Georgia, 2013).

[30] Schenk, T. RF Imperfections in High-rate Wireless Systems (Springer Netherlands, 2008).

[31] Roh, J. & Rao, B. Efficient feedback methods for MIMO channels based on parameterization. *IEEE Transactions on Wireless Communications* 6, 282–292 (2007).

[32] Porat, R., Ojard, E., Jindal, N., Fischer, M. & Erceg, V. Improved MU-MIMO performance for future 802.11 systems using differential feedback. In *UCSD Information Theory and Applications* Workshop, 1–5 (Catamaran Resort, San Diego, USA, 2013).

Chapter 4

Signal Processing by Generalized Receiver in DS-CDMA Wireless Communications Systems

Vyacheslav Tuzlukov

Additional information is available at the end of the chapter

http://dx.doi.org/10.5772/58990

1. Introduction

The additive and multiplicative noise exists forever in any wireless communication system. Quality and integrity of any wireless communication systems are defined and limited by statistical characteristics of the noise and interference, which are caused by an electromagnetic field of the environment.

The main characteristics of any wireless communication system are deteriorated as a result of the effect of the additive and multiplicative noise. The effect of addition of noise and interference to the signal generates an appearance of false information in the case of the additive noise. For this reason, the parameters of the received signal, which is an additive mixture of the signal, noise, and interference, differ from the parameters of the transmitted signal. Stochastic distortions of parameters in the transmitted signal, attributable to unforeseen changes in instantaneous values of the signal phase and amplitude as a function of time, can be considered as multiplicative noise. Under stimulus of the multiplicative noise, false information is a consequence of changed parameters of transmitted signals, for example, the parameters of transmitted signals are corrupted by the noise and interference. Thus, the impact of the additive noise and interference may be lowered by an increase in the signal-to-noise ratio (SNR). However, in the case of the multiplicative noise and interference, an increase in SNR does not produce any positive effects.

The main functional characteristics of any wireless communication systems are defined by an application area and are often specific for distinctive types of these systems. In the majority of cases, the main performance of any wireless communication systems are defined by some initial characteristics describing a quality of signal processing in the presence of noise: the precision of signal parameter measurement, the definition of resolution intervals of the signal parameters, and the probability of error.

The main idea is to use the **generalized approach to signal processing** (GASP) in noise in wireless communication systems [1-3]. The generalized approach is based on a seemingly abstract idea: the introduction of an additional noise source that does not carry any information about the signal and signal parameters in order to improve the qualitative performance of wireless communication systems. In other words, we compare statistical data defining the statistical parameters of the probability distribution densities (pdfs) of the observed input stochastic samples from two independent frequency time regions – a "yes" signal is possible in the first region and it is known a priori that a "no" signal is obtained in the second region. The proposed GASP allows us to formulate a decision-making rule based on the determination of **the jointly sufficient statistics of the mean and variance** of the likelihood function (or functional). Classical and modern signal processing theories allow us to define **only the mean** of the likelihood function (or functional). Additional information about the statistical characteristics of the likelihood function (or functional) leads us to better quality signal detection and definition of signal parameters in compared with the optimal signal processing algorithms of classical or modern theories.

Thus, for any wireless communication systems, we have to consider two problems – analysis and synthesis [8]. The first problem (analysis) – the problem to study a stimulus of the additive and multiplicative noise on the main principles and performance under the use of GASP – is an analysis of impact of the additive and multiplicative noise on the main characteristics of wireless communication systems, the receivers in which are constructed on the basis of GASP. This problem is very important in practice. This analysis allows us to define limitations on the use of wireless communication systems and to quantify the additive and multiplicative noise impact relative to other sources of interference present in these systems. If we are able to conclude that the presence of the additive and multiplicative noise is the main factor or one of the main factors limiting the performance of any wireless communication systems, then the second problem – the definition of structure and main parameters and characteristics of the generalized detector or receiver (GD or GR) under a dual stimulus of the additive and multiplicative noise – the problem of synthesis – arises.

GASP allows us to extend the well-known boundaries of the potential noise immunity set by classical and modern signal processing theories. Employment of wireless communication systems, the receivers of which are constructed on the basis of GASP, allows us to obtain high detection of signals and high accuracy of signal parameter definition with noise components present compared with that systems, the receivers of which are constructed on the basis of classical and modern signal processing theories. The optimal and asymptotic optimal signal processing algorithms of classical and modern theories, for signals with amplitude-frequency-phase structure characteristics that can be known and unknown a priori, are components of the signal processing algorithms that are designed on the basis of GASP.

In the proposed chapter, we would like to present and discuss the following aspects of GASP implemented in the direct-sequence code-division multiple access (DS-CDMA) wireless communication systems:

- The main theoretical statements and brief description;
- Signal processing in DS-CDMA wireless communication systems with optimal combining and partial cancellation;
- Signal processing in DS-CDMA wireless communication systems with frequency-selective channels;
- Signal processing in DS-CDMA downlink wireless communication systems with fading channels;
- Summary and discussion.

2. Main functioning principles under employment of GASP in DS-CDMA wireless communication systems

2.1. GR flowchart

The receiver in wireless communication system has an antenna array with the number of elements equal to M and each antenna array element receives N samples during the sensing time. The signal detection problem can be modeled as the conventional binary hypothesis test:

$$\begin{cases} \mathcal{H}_0 \Rightarrow z_i[k] = w_i[k], & i = 1,\dots,M;\ k = 0,\dots,N-1, \\ \mathcal{H}_1 \Rightarrow z_i[k] = h_i[k]\, s[k] + w_i[k], & i = 1,\dots,M;\ k = 0,\dots,N-1, \end{cases} \tag{1}$$

where $z_i[k]$ is the discrete-time received signal at the receiver input; $w_i[k]$ is the discrete-time circularly symmetric complex Gaussian noise with zero mean and variance σ^2, i.e. $w_i[k] \sim \mathcal{CN}(0,\sigma^2)$; $h_i[k]$ is the discrete-time channel coefficients obeying the circularly symmetric complex Gaussian distribution with zero mean and variance equal to σ_h^2, i.e., $h_i[k] \sim \mathcal{CN}(0,\sigma_h^2)$; and $s[k]$ is the discrete-time signal, i.e., the signal to be detected. We consider the same initial conditions with respect to $s[k]$ as in [4,5]. Throughout this chapter, the signal $s[k]$, the channel coefficients $h_i[k]$, and the noise $w_i[k]$ are independent between each other.

We define the $NM \times 1$ signal vector $\mathbf{Z}$ that collects all the observed signal samples during the sensing time:

$$\mathbf{Z} = \left[z_1[0],\ldots,z_M[0],\ldots,z_1[N-1],\ldots,z_M[N-1]\right]^T , \tag{2}$$

where T denotes a transpose. The data distribution in the complex matrix $\mathbf{Z}$ can be expressed as:

$$\mathbf{Z} \sim \begin{cases} \mathcal{CN}(0,\sigma^2\mathbf{I}) , & \Rightarrow \mathcal{H}_0 \\ \mathcal{CN}(0,E_s\sigma_h^2\mathbf{I}+\sigma^2\mathbf{I}) , & \Rightarrow \mathcal{H}_1 \end{cases} , \tag{3}$$

where E_s is the average signal energy at the receiver input, and $\mathbf{I}$ is the $MN \times MN$ identity matrix. We consider a situation when the signaling scheme is unknown (the receiver has a total freedom of choosing the signaling strategy) excepting a known power within the limits of the frequency band of interest. Thus, the receiver should be able to detect a presence of any possible signals satisfying the power and bandwidth constraints for robust detection of the signal $s[k]$ incoming at the receiver input in wireless communication systems.

The generalized receiver (GR) has been constructed based on the generalized approach to signal processing (GASP) in noise and discussed in numerous journal and conference papers and some monographs, namely, in [1–21]. GR is considered as a linear combination of the correlation detector that is optimal in the Neyman-Pearson criterion sense under detection of signals with known parameters and the energy detector that is optimal in the Neyman-Pearson criterion sense under detection of signals with unknown parameters. The main functioning principle of GR is a matching between the model signal generated by the local oscillator in GR and the information signal incoming at the GR input by whole range of parameters. In this case, the noise component of the GR correlation detector caused by interaction between the model signal generated by the local oscillator in GR and the input noise and the random component of the GR energy detector caused by interaction between the energy of incoming information signal and the input noise are cancelled in the statistical sense. This GR feature allows us to obtain the better detection performance in comparison with other classical receivers.

The specific feature of GASP is introduction of additional noise source that does not carry any information about the signal with the purpose to improve a qualitative signal detection performance. This additional noise can be considered as the reference noise without any information about the signal to be detected. The jointly sufficient statistics of the mean and variance of the likelihood function is obtained in the case of GASP implementation, while the classical and modern signal processing theories can deliver only a sufficient statistics of the mean or variance of the likelihood function (no the jointly sufficient statistics of the mean and variance of the likelihood function). Thus, the implementation of GASP allows us to obtain more information about the input process or received information signal. Owing to this fact, an implementation of the receivers constructed based on GASP basis allows us to improve the signal detection performance of wireless communication system in comparison with employment of other conventional receivers.

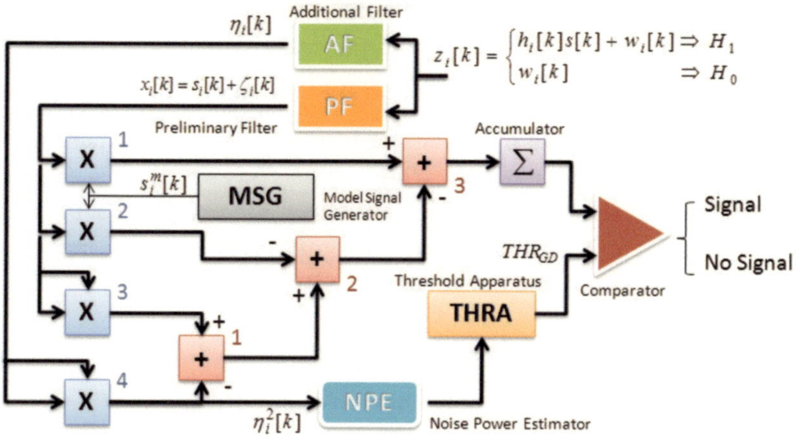

Figure 1. GR flowchart.

The GR flowchart is presented in Fig. 1. As we can see from Fig. 1, the GR consists of three channels:

- the correlation channel – the preliminary filter PF, multipliers 1 and 2, model signal generator MSG;
- the autocorrelation channel – the preliminary filter PF, the additional filter AF, multipliers 3 and 4, summator 1;
- the compensation channel – the summators 2 and 3 and accumulator 1.

As follows from Fig. 1, under the hypothesis $\mathcal{H}_1$ (a "yes" signal), the GR correlation channel generates the signal component $s_{\text{mod}_i}[k]s_i[k]$ caused by interaction between the model signal (the reference signal at the GR MSG output) and the incoming information signal and the noise component $s_{\text{mod}_i}[k]\xi_i[k]$ caused by interaction between the model signal $s_{\text{mod}_i}[k]$ and the noise $\xi_i[k]$ (the PF output). Under the hypothesis $\mathcal{H}_1$, the GR autocorrelation channel generates the information signal energy $s_i^2[k]$ and the random component $s_i[k]\xi_i[k]$ caused by interaction between the information signal $s_i[k]$ and the noise $\xi_i[k]$. The main purpose of the GR compensation channel is to cancel in the statistical sense the GR correlation channel noise component $s_{\text{mod}_i}[k]\xi_i[k]$ and the GR autocorrelation channel random component $s_i[k]\xi_i[k]$ between each other based on the same nature of the noise $\xi_i[k]$.

For description of the GR flowchart we consider the discrete-time processes without loss of any generality. Evidently, the cancelation between the GR correlation channel noise component $s_{\text{mod}_i}[k]\xi_i[k]$ and the GR autocorrelation channel random component $s_i[k]\xi_i[k]$ is possible only based on the same nature of the noise $\xi_i[k]$ satisfying the condition of equality between the signal model $s_{\text{mod}_i}[k]$ and incoming signal $s_i[k]$ over the whole range of parameters. The condition $s_{\text{mod}_i}[k] = s_i[k]$ is the **GR main functioning condition**. To satisfy this condition, we are able to define the incoming signal parameters. Naturally, in practice, signal parameters are random. How we can satisfy the GR main functioning condition and define the signal parameters in practice if there is no a priori information about the signal and there is an uncertainty in signal parameters, i.e. information signal parameters are stohastic, is discussed in detail in [1,3,21].

Under the hypothesis $\mathcal{H}_0$ (a "no" information signal), satisfying the GR main functioning condition $s_{\text{mod}_i}[k] = s_i[k]$, we obtain only the background noise $\eta_i^2[k] - \xi_i^2[k]$ at the GR output. Additionally, the practical implementation of the GR decision statistics requires an estimation of the noise variance σ_w^2 using the reference noise $\eta_i[k]$ at the AF output. AF is the reference noise source and the PF bandwidth is matched with the bandwidth of the information signal $s_i[k]$ to be detected. The threshold apparatus (THRA) device defines the GR threshold. PF and AF can be considered as the linear systems, for example, as the bandpass filters, with the impulse responses $h_{PF}[m]$ and $h_{AF}[m]$, respectively. For simplicity of analysis, we assume that these filters have the same amplitude-frequency characteristics or impulse responses by shape. Moreover, the AF central frequency is detuned with respect to the PF central frequency on such a value that the information signal cannot pass through the AF. Thus, the information signal and noise can be appeared at the PF output and the only noise is appeared at the AF output. If a value of detuning between the AF and PF central frequencies is more than $4 \text{ or } 5\Delta f_s$, where Δf_s is the signal bandwidth, the processes at the AF and PF outputs can be considered as the uncorrelated and independent processes and, in practice, under this condition, the coefficient of correlation between PF and AF output processes is not more than 0.05 that was confirmed experimentally in [22,23].

The processes at the AF and PF outputs present the input stochastic samples from two independent frequency-time regions. If the noise $w[k]$ at the PF and AF inputs is Gaussian, the noise at the PF and AF outputs is Gaussian, too, owing to the fact that PF and AF are the linear systems and we believe that these linear systems do not change the statistical parameters of the input process. Thus, the AF can be considered as a reference noise generator with a priori knowledge a "no" signal (the reference noise sample). A detailed discussion of the AF and PF can be found in [6,7]. The noise at the PF and AF outputs can be presented in the following form:

$$\begin{cases} w_{PF}[k] = \xi[k] = \sum_{m=-\infty}^{\infty} h_{PF}[m]w[k-m] \ , \\ w_{AF}[k] = \eta[k] = \sum_{m=-\infty}^{\infty} h_{AF}[m]w[k-m] \ . \end{cases} \tag{4}$$

Under the hypothesis $\mathcal{H}_1$, the signal at the PF output can be defined as $x_i[k] = s_i[k] + \xi_i[k]$ (see Fig. 1), where $\xi_i[k]$ is the observed noise at the PF output and $s_i[k] = h_i[k] \times s[k]$; $h_i[k]$ are the channel coefficients indicated here only in a general case. Under the hypothesis $\mathcal{H}_0$ and for all i and k, the process $x_i[k] = \xi_i[k]$ at the PF output is subjected to the complex Gaussian distribution and can be considered as the independent and identically distributed (i.i.d.) process. The process at the AF output is the reference noise $\eta_i[k]$ with the same statistical parameters as the noise $\xi_i[k]$ in the ideal case (we make this assumption for simplicity). In practice, the statistical parameters of the noise $\xi_i[k]$ and $\eta_i[k]$ are different.

The decision statistics at the GR output presented in [1,3] is extended to the case of antenna array employment when an adoption of multiple antennas and antenna arrays is effective to mitigate the negative attenuation and fading effects [4,5]. The GR decision statistics can be presented in the following form:

$$T_{GR}(\mathbf{X}) = \sum_{k=0}^{N-1}\sum_{i=1}^{M} 2x_i[k]s_{\text{mod}_i}[k] - \sum_{k=0}^{N-1}\sum_{i=1}^{M} x_i^2[k] + \sum_{k=0}^{N-1}\sum_{i=1}^{M} \eta_i^2[k] \underset{\mathcal{H}_0}{\overset{\mathcal{H}_1}{\gtrless}} THR_{GR}, \tag{5}$$

where $\mathbf{X} = [x_1[0],\ldots,x_M[0],\ldots,x_1[N-1],\ldots,x_M[N-1]]^T$ is the vector of random process at the PF output, and THR_{GR} is the GR detection threshold. We can rewrite (5) using the vector form:

$$T_{GR}(\mathbf{X}) = 2\mathbf{S}_{\text{mod}}\mathbf{X} - \mathbf{X}^2 + \boldsymbol{\eta}^2 \underset{\mathcal{H}_0}{\overset{\mathcal{H}_1}{\gtrless}} THR_{GR}, \tag{6}$$

where $\mathbf{X} = [\mathbf{x}(0),..., \mathbf{x}(N-1)]$ is the $M \times 1$ vector of the random process at the PF output with elements defined as $\mathbf{x}[k] = [x_1[k],\ldots,x_M[k]]^T$; $\mathbf{S}_{\text{mod}} = [\mathbf{s}_{\text{mod}}(0),\ldots,\mathbf{s}_{\text{mod}}(N-1)]$ is the $M \times 1$ vector of the process at the MSG output with the elements defined as $\mathbf{s}_{\text{mod}}[k] = [s_{\text{mod}_1}[k],\ldots,s_{\text{mod}_M}[k]]^T$; $\boldsymbol{\eta} = [\boldsymbol{\eta}(0),\ldots,\boldsymbol{\eta}(N-1)]$ is the $M \times 1$ vector of the random process at the AF output with the elements defined as $\boldsymbol{\eta}[k] = [\eta_1[k],\ldots,\eta_M[k]]^T$ and THR_{GR} is the GR detection threshold.

According to GASP and GR structure shown in Fig. 1, the GR test statistics takes the following form under the hypotheses $\mathcal{H}_1$ and $\mathcal{H}_0$, respectively:

$$T_{GR}(\mathbf{X}) = \begin{cases} \sum_{k=0}^{N-1}\sum_{i=1}^{M} s_i^2[k] + \sum_{k=0}^{N-1}\sum_{i=1}^{M} \eta_i^2[k] - \sum_{k=0}^{N-1}\sum_{i=1}^{M} \xi_i^2[k] & \Rightarrow \mathcal{H}_1 , \\ \sum_{k=0}^{N-1}\sum_{i=1}^{M} \eta_i^2[k] - \sum_{k=0}^{N-1}\sum_{i=1}^{M} \xi_i^2[k] & \Rightarrow \mathcal{H}_0 . \end{cases} \quad (7)$$

The term $\sum_{k=0}^{N-1}\sum_{i=1}^{M} s_i^2[k] = E_s$ is the average signal energy and the term $\sum_{k=0}^{N-1}\sum_{i=1}^{M} \eta_i^2[k] - \sum_{k=0}^{N-1}\sum_{i=1}^{M} \xi_i^2[k]$ presents the background noise at the GR output that is a difference between the noise power at the PF and AF outputs. It is important to mention that the main GR functioning condition is the equality between parameters of the model signal $s_{\text{mod}_i}[k]$ and the signal $s_i[k]$, i.e. $s_{\text{mod}_i}[k] = s_i[k]$ over all range of parameters and, in particular, by amplitude. How we can satisfy this condition in practice is discussed in detail in [1] and [3, Chapter 7, pp 611-621] when there is no a priori information about the signal $s_i[k]$. This condition is essential for complete compensation in the statistical sense between the noise component of the correlation channel $2s_{\text{mod}_i}[k]\xi_i[k]$, the GR correlation channel, caused by interaction between the model signal $s_{\text{mod}_i}[k]$ and noise $\xi_i[k]$, and the random component of the GR autocorrelation channel $2s_i[k]\xi_i[k]$, the GR ED, caused by interaction between the signal $s_i[k]$ and noise $\xi_i[k]$ [1] and [6, Chapter 3]. The complete matching between the model signal $s_{\text{mod}_i}[k]$ and the incoming signal $s_i[k]$, especially by amplitude, is a very hard problem in practice and only in the ideal case the complete matching is possible. How the GR detection performance can be deteriorated under mismatching between the model signal $s_{\text{mod}_i}[k]$ and the incoming signal $s_i[k]$ is discussed in [24]. Additionally, a practical implementation of the GR decision statistics requires an estimation of the noise variance σ_w^2 using the reference noise $\eta_i[k]$ at the AF output.

The mean $m_{\mathcal{H}_0}^{GR}$ and the variance $Var_{\mathcal{H}_0}^{GR}$ of the test statistics $T_{GR}(\mathbf{X})$ under the hypothesis $\mathcal{H}_0$ are given in the following form [6, Chapter 3]:

$$\begin{cases} m_{\mathcal{H}_0}^{GR} = \mathrm{E}\left[T_{GR}(\mathbf{X}) \middle| \mathcal{H}_0\right] = 0 , \\ Var_{\mathcal{H}_0}^{GR} = Var\left[T_{GR}(\mathbf{X}) \middle| \mathcal{H}_0\right] = 4NM\sigma^4 . \end{cases} \quad (8)$$

2.2. MGF of the GR partial test statistics $T_{GR}(\mathbf{X}_k)$

To complete consideration of the GR main functioning principles the moment generating function (MGF) definition of the GR partial test statistics $T_{GR}(\mathbf{X}_k)$ under the main GR functioning condition $s_{\text{mod}_i}[k] = s_i[k]$ and the hypothesis $\mathcal{H}_1$ given by

$$T_{GR}(\mathbf{X}_k) = \sum_{i=1}^{M} s_i^2[k] + \sum_{i=1}^{M} \eta_i^2[k] - \sum_{i=1}^{M} \xi_i^2[k] \tag{9}$$

is required. We say that the random variable x has a chi-square distribution with υ degree of freedom if its probability density function (pdf) is determined as

$$p(x) = cx^{0.5\upsilon-1}\exp(-0.5x)\ , \tag{10}$$

where c is a constant given by [25]

$$c = \frac{1}{2^{0.5\upsilon}\Gamma(0.5\upsilon)}\ , \tag{11}$$

$\Gamma(\cdot)$ is the gamma function. The MGF general form for the chi-square distributed random variable x is given by [25]

$$\mathcal{M}_x(l) = \mathrm{E}[\exp(lx)] = \int_{-\infty}^{\infty} \exp(lx)p(x)dx = c\int_{0}^{\infty} \exp(lx)x^{0.5\upsilon-1}\exp(-0.5x)\,dx\ , \tag{12}$$

where $\mathrm{E}[\cdot]$ is the mathematical expectation. At $\upsilon=1$, the constant

$$c = \frac{1}{2^{0.5}\Gamma(0.5\upsilon)} = \frac{1}{\sqrt{2\pi}}\ . \tag{13}$$

Assume that $z_{1_i}[k] = \xi_i^2[k]$ and $z_{2_i}[k] = \eta_i^2[k]$. The pdfs for the random variables z_{1_i} and z_{2_i} are defined by the chi square χ^2 distribution law with one degree of freedom [3]:

$$p(z_{1i}) = \frac{1}{\sqrt{2\pi\ z_{1i}}\ \sigma_w}\exp\left\{-\frac{z_{1i}}{2\sigma_w^2}\right\}\ ,\ z_{1i} > 0\ , \tag{14}$$

$$p(z_{2i}) = \frac{1}{\sqrt{2\pi\ z_{2i}}\ \sigma_w}\exp\left\{-\frac{z_{2i}}{2\sigma_w^2}\right\}\ ,\ z_{2i} > 0\ . \tag{15}$$

Introduce a new random variable $z_i = z_{1_i} - z_{2_i}$. The MGF of the random variable z_i is given using the following formula:

$$\begin{aligned}\mathcal{M}_{z_i}(l) &= \mathrm{E}[\exp(lz_i)] = \mathrm{E}\{\exp[l(z_{1_i} - z_{2i})]\} = \mathrm{E}[\exp(lz_{1_i})\exp(-lz_{2_i})] \\ &= \mathrm{E}[\exp(lz_{1_i})]\mathrm{E}[\exp(-lz_{2_i})] = \mathcal{M}_{z_{1i}}(l)\mathcal{M}_{z_{2i}}(-l).\end{aligned} \tag{16}$$

The MGF of the random variable z_1 is defined in the following form:

$$\mathcal{M}_{z_{1_i}}(l) = \frac{1}{\sqrt{2\pi}\,\sigma_w}\int_0^\infty \exp(lz_{1_i}) z_{1_i}^{-0.5} \exp\left\{-\frac{z_{1_i}}{2\sigma_w^2}\right\} dz_{1_i} = \frac{1}{\sqrt{2\pi}\,\sigma_w}\int_0^\infty z_{1_i}^{-0.5} \exp\left\{-\left(\frac{1}{2\sigma_w^2} - l\right) z_{1_i}\right\} dz_{1_i}\,. \quad (17)$$

Introducing the variable $g_i = \left(0.5\sigma_w^{-2} - l\right) z_{1_i}$, we can obtain:

$$\begin{aligned}\mathcal{M}_{z_{1i}}(l) &= \frac{2\sigma_w^2}{\sqrt{2\pi}\sigma_w}\int_0^\infty \left[2\sigma_w^2(1-2\sigma_w^2 l)^{-1}\right]^{-\frac{1}{2}} g_i^{-0.5}\exp(-g_i)(1-2\sigma_w^2 l)^{-1} dg_i \\ &= \frac{1}{\sqrt{2\pi}\sigma_w}\left[2\sigma_w^2(1-2\sigma_w^2 l)^{-1}\right]^{-\frac{1}{2}}\int_0^\infty g_i^{-0.5}\exp(-g_i)dg_i = \left[\pi(1-2\sigma_w^2 l)\right]^{-\frac{1}{2}}\int_0^\infty g_i^{-0.5}\exp(-g_i)dg_i.\end{aligned} \quad (18)$$

Based on the definition of the gamma function [26]

$$\Gamma(x) = \int_0^\infty l^{x-1}\exp(-l)dl\ , \quad (19)$$

we obtain that

$$\int_0^\infty g_i^{-0.5}\exp(-g_i)dg_i = \Gamma(0.5) = \sqrt{\pi}. \quad (20)$$

Finally, the MGF of the random variable z_{1_i} is defined as

$$\mathcal{M}_{z_{1_i}}(l) = \sqrt{\pi}\left[\pi(1-2\sigma_w^2 l)\right]^{-\frac{1}{2}} = (1-2\sigma_w^2 l)^{-\frac{1}{2}}\,. \quad (21)$$

The mean and the variance of the random variables z_{1_i} can be determined in the following form:

$$\mathrm{E}[z_{1i}] = \left.\frac{\partial \mathcal{M}_{z_{1i}}(l)}{\partial l}\right|_{l=0} = \sigma_w^2\ , \quad (22)$$

$$Var[z_{1_i}] = \mathrm{E}[z_{1i}^2] - \mathrm{E}[z_{1_i}]^2 = \left.\frac{\partial^2 \mathcal{M}_{z_{1_i}}(l)}{\partial^2 l}\right|_{l=0} - \mathrm{E}[z_{1_i}]^2 = 3\sigma_w^4 - \sigma_w^4 = 2\sigma_w^4\ . \quad (23)$$

http://dx.doi.org/10.5772/58990

By the analogous way, we can find that the MGF of the random variable z_{2_i} takes the following form:

$$\mathcal{M}_{z_{2_i}}(-l) = (1 + 2\sigma_w^2 l)^{-0.5} \ . \tag{24}$$

Since $\{s_i[k]\}_{i=1}^M$ are spatially correlated for i-th antenna array elements, and according to [28], the MGF of $\sum_{i=1}^M s_i^2[k]$ is defined as

$$\mathcal{M}_{\sum_{i=1}^M s_i^2[k]}(l) = \prod_{i=1}^M (1 - E_s \sigma_h^2 \beta_i l)^{-1}, \tag{25}$$

where β_i is the eigenvalue of the i-th spatial channel of the correlation matrix $\mathbf{C}$ given by (2). Based on (21), (24), and (25), the MGF of the GD partial decision statistics $T_{GR}(\mathbf{X}_k)$ is determined in the following form:

$$\begin{aligned}\mathcal{M}_{T_{GR}(\mathbf{X}_k)}(l) &== \prod_{i=1}^M [1 - E_s \sigma_h^2 \beta_i l]^{-1} \prod_{i=1}^M \mathcal{M}_{z_{1_i}}(l) \prod_{i=1}^M \mathcal{M}_{z_{2_i}}(-l) = \prod_{i=1}^M [1 - E_s \sigma_h^2 \beta_i l]^{-1} \prod_{i=1}^M (1 - 2\sigma_w^2 l)^{-0.5} \prod_{i=1}^M (1 + 2\sigma_w^2 l)^{-0.5} \\ &= \prod_{i=1}^M [1 - E_s \sigma_h^2 \beta_i l]^{-1} (1 - 2\sigma_w^2 l)^{-0.5M} (1 + 2\sigma_w^2 l)^{-0.5M} = (1 - 4\sigma_w^4 l^2)^{-0.5M} \prod_{i=1}^M [1 - E_s \sigma_h^2 \beta_i l]^{-1} \ .\end{aligned} \tag{26}$$

Based on (26) and taking into consideration results discussed in [27], the mean and the variance of the test statistics $T_{GR}(\mathbf{X})$ under the hypothesis $\mathcal{H}_1$ take the following form, respectively:

$$m_{\mathcal{H}_1}^{GR} = \mathrm{E}\left[T_{GR}(\mathbf{X}_k) \middle| \mathcal{H}_1\right] = NM(E_s \sigma_h^2 + 2\sigma_w^2) \ , \tag{27}$$

$$Var_{\mathcal{H}_1}^{GR} = Var\left[T_{GR}(\mathbf{X}_k) \middle| \mathcal{H}_1\right] = N\left[\sum_{i=1}^M E_s^2 \sigma_h^4 \beta_i^2 + 4M\sigma_w^4\right]. \tag{28}$$

3. Signal processing with optimal combining and partial cancellation

3.1. Brief review

In this section, we investigate the generalized receiver (GR) under the quadrature subbranch hybrid selection/maximal-ratio combining (HS/MRC) for 1-D modulations in multipath fading channel and compare its symbol error rate (SER) performance with that of the traditional HS/MRC scheme discussed in [29,30]. It is well known that the HS/MRC receiver selects the L strongest signals from N available diversity branches and coherently combines them. In traditional HS/MRC scheme, the strongest L signals are selected according to signal-envelope amplitude [29–35]. However, some receiver implementations recover directly the in-phase and quadrature components of the received branch signals. Furthermore, the optimal maximum likelihood estimation (MLE) of the phase of a diversity branch signal is implemented by first estimating the in-phase and quadrature branch signal components and obtaining the signal phase as a derived quantity [36,37]. Other channel-estimation procedures also operate by first estimating the in-phase and quadrature branch signal components [38–41]. Thus, rather than N available signals, there are $2N$ available quadrature branch signal components for combining. In general, the largest $2L$ of these $2N$ quadrature branch signal components will not be the same as the $2L$ quadrature branch signal components of the L branch signals having the largest signal envelopes.

In this section, we investigate how much improvement in performance can be achieved employing the GR with modified HS/MRC, namely, with the quadrature subbranch HS/MRC and HS/MRC schemes, instead of the conventional HS/MRC combining scheme for 1-D signal modulations in multipath fading channel. At GR discussed in [42], the N diversity branches are split into $2N$ in-phase and quadrature subbranches. Then the GR with HS/MRC scheme is applied to these $2N$ subbranches. Obtained results show the better performance is achieved by this quadrature subbranch HS/MRC scheme in comparison with the traditional HS/MRC scheme for the same value of average signal-to-noise ratio (SNR) per diversity branch.

Another problem discussed is the problem of partial cancellation factor (PCF) in DS-CDMA wireless communication system with multipath fading channel. It is well known that the multiple access interference (MAI) can be efficiently estimated by the partial parallel interference cancellation (PPIC) procedure and then partially be cancelled out of the received signal on a stage-by-stage basis for DS-CDMA wireless communication system [43]. To ensure a high-quality performance, PCF for each PPIC stage needs to be chosen appropriately, where the PCF should be increased as the reliability of the MAI estimates improves. There are some papers on the selection of the PCF for a receiver based on the PPIC. In [44–46], formulas for the optimal PCF were derived through straightforward analysis based on soft decisions. In contrast, it is very difficult to obtain the optimal PCF for a receiver based on PPIC with hard decisions owing to their nonlinear character. One common approach to solve the nonlinear problem is to select an arbitrary PCF for the first stage and then increase the value for each successive stage, since the MAI estimates may become more reliable in later stages [43, 47, 48]. This approach is simple, but it might not provide satisfactory performance.

In this section, we use the Price's theorem [49, 50] to derive a range of the optimal PCF for the first stage in PPIC of DS-CDMA wireless communication system with multipath fading channel employing GR based on GASP [1–3], where the lower and upper boundary values of the PCF can be explicitly calculated from the processing gain and the number of users of DS-CDMA wireless communication system in the case of periodic code scenario. Computer simulation shows that, using the average of these two boundary values as the PCF for the first stage, we are able to reach the bit error rate (BER) performance that is very close to the potentially achieved one [51] and surpasses the BER performance of the real PCF for DS-CDMA wireless communication systems discussed, for example, in [43].

With this result, a reasonable PCF can quickly be determined without using any time-consuming Monte Carlo simulations. It is worth mentioning that the two-stage GR considered in [52] based on the PPIC using the proposed PCF at the first stage achieves the BER performance comparable to that of the three-stage GR based on the PPIC using an arbitrary PCF at the first stage. In other words, at the same BER performance, the proposed approach for selecting the PCF can reduce the GR complexity based on the PPIC. The PCF selection approach is derived for multipath fading channel cases discussed in [42, 53].

In this section, we describe the multipath fading channel model and provide system models for selection/maximal ratio combining and synchronous DS-CDMA wireless communication systems; carry out the performance analysis obtaining a symbol error rate expression in the closed-form and define a marginal moment generating function of SNR per symbol of a single quadrature branch; determine the lower and upper PCF bounds based on the processing gain N and the number of users K under multipath fading channel model in DS-CDMA wireless communication systems employing GR; discuss simulation results; and make some conclusions.

3.2. System model

3.2.1. Multipath fading channel model

Let the transfer function for user k's channel be

$$W_k(Z) = \sum_{i=1}^{M} \alpha_{k,i} Z^{-\tau_{k,i}}. \tag{29}$$

As we can see from (29), the number of paths is M and the channel power and delay for i-th channel path are $\alpha_{k,i}$ and $\tau_{k,i}$, respectively.

We use two vectors to represent these parameters:

$$\boldsymbol{\tau}_k = [\tau_{k,1}, \tau_{k,2}, \ldots, \tau_{k,L}]^T \tag{30}$$

and

$$\boldsymbol{\alpha}_k = [\alpha_{k,1}, \alpha_{k,2}, \ldots, \alpha_{k,L}]^T. \tag{31}$$

Let

$$\tau_{k,1} \le \tau_{k,2} \le \cdots \le \tau_{k,L} \tag{32}$$

and the channel power is normalized

$$\sum_{i=1}^{L} \alpha_{k,i}^2 = 1. \tag{33}$$

Without loss of generality, we may assume that $\tau_{k,1} = 0$ for each user and L is the maximum possible number of paths. When a user's path number, say M_1, is less than M, we can let all the elements in $\tau_{k,i}$ and $\alpha_{k,i}$ be zero if the following condition is satisfied

$$M_1 + 1 \le i \le M. \tag{34}$$

We may also assume that the maximum delay is much smaller than the processing gain N [46]. Before our formulation, we first define a $(2N-1)\times L$ composite signature matrix $\mathbf{A}_k$ in the following form

$$A_k = [\tilde{\mathbf{a}}_{k,1}, \tilde{\mathbf{a}}_{k,2}, \ldots, \tilde{\mathbf{a}}_{k,L}] \tag{35}$$

where $\tilde{\mathbf{a}}_{k,i}$ is a vector containing i-th delayed spreading code for user k. It is defined as

$$\tilde{\mathbf{a}}_{k,i} = [\overbrace{0,\ldots,0}^{\tau_{k,i}}, \boldsymbol{a}_k^T, \overbrace{0,\ldots,0}^{N-\tau_{k,i}-1},]^T. \tag{36}$$

Since a multipath fading channel is involved, the current received bit signal will be interfered by previous bit signal. As mentioned above, the maximum path delay is much smaller than the processing gain. The interference will not be severe and for simplicity, we may ignore this effect. Let us denote the channel gain for multipath fading as

$$\mathbf{h}_k = \boldsymbol{\alpha}_k \mathbf{A}_k. \tag{37}$$

3.2.2. *Selection/maximal-ratio combining*

We assume that there are N diversity branches experiencing slow and flat Rayleigh fading, and all of the fading processes are independent and identically distributed (i.i.d.). During analysis, we consider only the hypothesis H_1 "a yes" signal in the input stochastic process. Then the equivalent received baseband signal for the k-th diversity branch takes the following form:

$$x_k(t) = h_k(t)s(t-\tau_k) + w_k(t), \quad k = 1,\ldots,N, \tag{38}$$

where $s(t-\tau_k)$ is a 1-D baseband transmitted signal that without loss of generality, is assumed to be real, $h_k(t)$ is the complex channel gain for the k-th branch subjected to Rayleigh fading, τ_k is the propagation delay along the k-th path of received signal, and $w_k(t)$ is the zero-mean complex AWGN with two-sided power spectral density $N_0/2$ with the dimension W/Hz. At GR front end, for each diversity branch, the received signal is split into its in-phase and quadrature signal components. Then, the conventional HS/MRC scheme is applied over all of these quadrature branches, as shown in Fig.2.

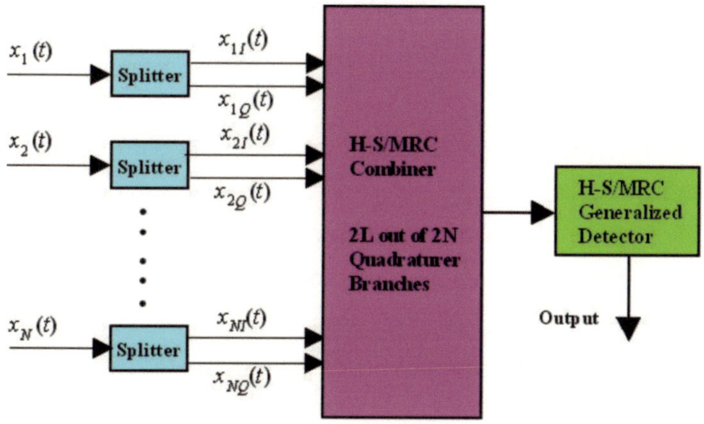

Figure 2. Block diagram receiver based on GR with quadrature subbranch HS/MRC and HS/MRC schemes.

We can present $h_k(t)$ given by (29)–(37) as i.i.d. complex Gaussian random variables assuming that each of the L branches experiences the slow and flat Rayleigh fading

$$h_k(t) = \alpha_k(t)\exp\{-j\phi_k(t)\} = \alpha_k \exp\{-j\phi_k\}, \tag{39}$$

where α_k is a Rayleigh random variable and φ_k is a random variable uniformly distributed within the limits of the interval $[0,2\pi)$. Owing to the fact that the fade amplitudes are Rayleigh distributed, we can present $h_k(t)$ as

$$h_k(t) = h_{kI}(t) + jh_{kQ}(t) \tag{40}$$

and $w_k(t)$ as

$$w_k(t) = w_{kI}(t) + jw_{kQ}(t). \tag{41}$$

The in-phase signal component $x_{kI}(t)$ and quadrature signal component $x_{kQ}(t)$ of the received signal $x_k(t)$ are given by

$$x_{kI}(t) = h_k(t)a(t-\tau_k) + w_{kI}(t), \tag{42}$$

$$x_{kQ}(t) = h_{kQ}(t)a(t-\tau_k) + w_{kQ}(t). \tag{43}$$

Since $h_k(t), (k=1,\ldots,K)$ are subjected to i.i.d. Rayleigh fading, we can assume that the in-phase $h_{kI}(t)$ and quadrature $h_{kQ}(t)$ channel gain components are independent zero-mean Gaussian random variables with the same variance [41]

$$\sigma_h^2 = 0.5E[\,|\,h_k^2(t)\,|\,], \tag{44}$$

where $E[\cdot]$ is the mathematical expectation. Further, the in-phase $w_{kI}(t)$ and quadrature $w_{kQ}(t)$ noise components are also the independent zero-mean Gaussian random processes, each with two-sided power spectral density $0.5N_0$ with the dimension W/Hz [36]. Due to the independence of the in-phase $h_{kI}(t)$ and quadrature and quadrature $h_{kQ}(t)$ channel gain components and the in-phase $w_{kI}(t)$ and quadrature $w_{kQ}(t)$ noise components, the 2N quadrature branch received signal components conditioned on the transmitted signal are i.i.d.

We can reorganize the in-phase and quadrature components of the channel gains h_k and Gaussian noise $w_k(t)$ when $k=1,\ldots,N$ as g_k and v_k, given, respectively by

$$g_k(t) = \begin{cases} h_{kI}(t), & k=1,\ldots,N \\ h_{(k-N)Q}(t), & k=N+1,\ldots,2N; \end{cases} \tag{45}$$

$$v_k(t) = \begin{cases} w_{kI}(t), & k=1,\ldots,N \\ w_{(k-N)Q}(t), & k=N+1,\ldots,2N. \end{cases} \tag{46}$$

The GR output with quadrature subbranch HS/MRC and HS/MRC schemes according to GASP [2, 3, 6–9] is given by:

$$Z_{QBHS/MRC}^{GR}(t) = s^2(t)\sum_{k=1}^{2N} b_k^2 g_k^2(t) + \sum_{k=1}^{2N} b_k^2 g_k^2(t)\left[v_{k_{AF}}^2(t) - v_{k_{PF}}^2(t)\right], \tag{47}$$

where $v_{k_{AF}}^2(t) - v_{k_{PF}}^2(t)$ is the background noise forming at the GR output for the *k*-th branch; $b_k \in \{0,1\}$ and 2*L* of the b_k equal 1.

3.2.3. Synchronous DS-CDMA wireless communication system

Consider a synchronous DS-CDMA system employing the GR with K users, the processing gain N, the number of frame L, the chip duration T_c, the bit duration $T_b = \frac{NT_c}{R}$ with information bit encoding rate R. The signature waveform of the user k is given by

$$a_k(t) = \sum_{i=1}^{N} a_{ki} p_{T_c}(t - iT_c), \tag{48}$$

where $\{a_{k1}, a_{k2}, ..., a_{kN}\}$ is a random spreading code with each element taking value on $\pm 1/\sqrt{N}$ equiprobably, $p_{T_c}(t)$ is the unit amplitude rectangular pulse with duration T_c. The baseband signal transmitted by the user k is given by

$$s_k(t) = A_k(t) \sum_{i=1}^{L} b_{k,i} a_k(t - iT_b), \tag{49}$$

where $A_k(t)$ is the transmitted signal amplitude of the user k. The following form can present the received baseband signal:

$$x(t) = \sum_{k=1}^{K} h_k(t) s_k(t) + w(t) = \sum_{k=1}^{K} \sum_{i=1}^{L} S_k(t) b_{k,i} a_k(t - iT_b) + w(t) \ , \quad t \in [0, T_b] \tag{50}$$

where taking into account (29)–(37) and (39) and as it was shown in [13]

$$S_k(t) = h_k(t) A_k(t) = \alpha_k^2 A_k(t) \tag{51}$$

is the received signal amplitude envelope for the user k, $w(t)$ is the complex Gaussian noise with zero mean with

$$E[w_k(t)(w_j(t))^*] = \begin{cases} 2\alpha_k^2 \sigma_w^2 \ , & \text{if} \quad j = k \\ 2\alpha_k^2 \sigma_w^2 \rho_{kj} \ , & \text{if} \quad j \neq k \end{cases}, \tag{52}$$

ρ_{kj} is the coefficient of correlation. Using GR based on the multistage PPIC for DS-CDMA systems and assuming the user k is the desired user, we can express the corresponding GR output according to GASP and the main functioning condition of GR expressed by $s_{\text{mod}_i}[k] = s_i[k]$ as the first stage of the PPIC GR:

$$Z_k(t) = \int_0^{T_b} [2x_k(t)s_k^*(t-\tau_k) - x_k(t)x_k(t-\tau_k)]dt + \int_0^{T_b} \alpha_k^2[w_{k_{AF}}(t)w_{k_{AF}}(t-\tau_k) - w_{k_{PF}}(t)w_{k_{PF}}(t-\tau_k)]dt, \quad (53)$$

where $s_k^*(t)$ is the model of the signal transmitted by the user k ($s_{\text{mod}_i}[k] = s_i[k]$); τ_k is the delay factor that can be neglected for simplicity of analysis. For this case, we have

$$Z_k = S_k(t)b_k + \sum_{j=1, j\neq k}^{K} S_j(t)b_j\rho_{kj} + \zeta_k(t) = S_k(t)b_k + I_k(t) + \zeta_k(t) = h_k(t)A_k(t)b_k + I_k(t) + \zeta_k(t), \quad (54)$$

where the first term in (54) is the desired signal;

$$\rho_{kj} = \int_0^{T_b} s_k(t)s_j(t)dt \quad (55)$$

is the coefficient of correlation between signature waveforms of the k-th and j-th users; the third term in (54)

$$\zeta_k(t) = \int_0^{T_b} \alpha_k^2[w_{k_{AF}}^2(t) - w_{k_{PF}}^2(t)]\, dt \quad (56)$$

is the total noise component at the GR output; and the second term in (54)

$$I_k = \sum_{j=1, j\neq k}^{K} S_j b_j \rho_{kj} = \sum_{j=1, j\neq k}^{K} h_j A_j b_j \rho_{kj} = \sum_{j=1, j\neq k}^{K} \alpha_j^2 A_j b_j \rho_{kj} \quad (57)$$

is the MAI. The conventional GR makes a decision based on Z_k. Thus, MAI is treated as another noise source. When the number of users is large, MAI will seriously degrade the system performance. GR with partial interference cancellation, being a multiuser detection scheme [31], is proposed to alleviate this problem.

Denoting the soft and hard decisions at the GR output for the user k by

$$\tilde{b}_k^{(0)} = Z_k \text{ and } \hat{b}_k^{(0)} = sgn(Z_k) \quad (58)$$

respectively, the output of the GR with the first PPIC stage with a partial cancellation factor equal to p_1 can be written by [43]

$$\tilde{b}_k^{(1)} = p_1(Z_k - \hat{I}_k) + (1 - p_1)\tilde{b}_k^{(0)} = Z_k - p_1\hat{I}_k, \tag{59}$$

where $\tilde{b}_k^{(1)}$ denotes the soft decision of user k at the GR output with the first stage of PPIC and

$$\hat{I}_k = \sum_{j=1, j\neq k}^{K} S_j \hat{b}_j^{(0)} \rho_{kj} = \sum_{j=1, j\neq k}^{K} h_j A_j \hat{b}_j^{(0)} \rho_{kj} = \sum_{j=1, j\neq k}^{K} \alpha_j^2 A_j \hat{b}_j^{(0)} \rho_{kj} \tag{60}$$

is the estimated MAI using a hard decision.

3.3. Performance analysis

3.3.1. Symbol error rate expression

Let q_k denote the instantaneous SNR per symbol of the k-th quadrature branch $(k = 1,\ldots,2N)$ at the GR output under quadrature subbranch HS/MRC and HS/MRC schemes. In line with [3, 46] and (29)–(37) and (39), the instantaneous SNR q_k can be defined in the following form:

$$q_k = \frac{E_b \alpha_k^2}{2\sigma_w^2}, \tag{61}$$

where E_b is the average symbol energy of the transmitted signal $s(t)$.

Assume that we choose $2L\,(1 \le L \le N)$ quadrature branched out of the $2N$ branches. Then, the SNR per symbol at the GR output under quadrature subbranch HS/MRC and HS/MRC schemes may be presented as

$$q_{QBHS/MRC} = \sum_{k=1}^{2L} q_{(k)}, \tag{62}$$

where $q_{(k)}$ are the ordered instantaneous SNRs q_k and satisfy the following condition

$$q_{(1)} \ge q_{(2)} \ge \cdots \ge q_{(2N)}. \tag{63}$$

When $L = N$, we obtain the MRC, as expected. Using the MGF method discussed in [33, 41], SER of M-ary pulse amplitude modulation (PAM) system conditioned on $q_{QBHS/MRC}$ is given by

$$P_s(q_{QBHS/MRC}) = \frac{2(M-1)}{M\,\pi}\int_0^{0.5\pi} \exp\Big(-\frac{g_{M-PAM}}{\sin^2\theta} q_{QBHS/MRC}\Big)d\theta, \tag{64}$$

where

$$g_{M-PAM} = \frac{3}{M^2-1}. \tag{65}$$

Averaging (64) over $q_{QBHS/MRC}$ the SER of M-ary PAM system is determined in the following form:

$$P_s = \frac{2(M-1)}{M\,\pi}\int_0^{0.5\pi}\int_0^{\infty} \exp\Big(-\frac{g_{M-PAM}}{\sin^2\theta} q\Big) f_{q_{QBHS/MRC}}(q)dqd\theta = \frac{2(M-1)}{M\,\pi}\int_0^{0.5\pi} \varphi_{q_{QBHS/MRC}}\Big(-\frac{g_{M-PAM}}{\sin^2\theta}\Big)d\theta, \tag{66}$$

where

$$\varphi_q(s) = E_q\{\exp(sq)\} \tag{67}$$

is the MGF of random variable q, $E_q\{\cdot\}$ is the mathematical expectation of MGF with respect to SNR per symbol.

A finite-limit integral for the Gaussian Q-function, which is convenient for numerical integrations, is given by [54]

$$Q(x) = \begin{cases} \dfrac{1}{\pi}\displaystyle\int_0^{0.5\pi} \exp\left\{-\frac{x^2}{2\sin^2\theta}\right\}d\theta\,, & x \ge 0 \\ 1-\dfrac{1}{\pi}\displaystyle\int_0^{0.5\pi} \exp\left\{-\frac{x^2}{2\sin^2\theta}\right\}d\theta\,. & x < 0 \end{cases} \tag{68}$$

The error function can be related to the Gaussian Q-function by

$$erf(x) = \frac{2}{\sqrt{\pi}}\int_0^x \exp(-t^2)dt = 1 - 2Q(\sqrt{2}x). \tag{69}$$

The complementary error function is defined as $erfc(x) = 1 - erf(x)$ so that

$$Q(x) = \frac{1}{2}erfc\left(\frac{x}{\sqrt{2}}\right) \quad \text{or} \quad erfc(x) = 2Q(\sqrt{2}x), \tag{70}$$

http://dx.doi.org/10.5772/58990

which is convenient for computing values using MATLAB since **erfc** is a subprogram in MATLAB but the Gaussian *Q*-function is not (unless you have a *Communications Toolbox*). Note that the Gaussian *Q*-function is the tabulated function.

Now, let us compare (64) and (68) to obtain the closed form expression for the SER of *M*-ary PAM wireless communication system employing the GR with quadrature subbranch HS/MRC and HS/MRC schemes. We can easily see that taking into account (44), (45), (61), (62), and (65), the SER of *M*-ary PAM system employing the GR with quadrature subbranch HS/MRC and HS/MRC schemes can be defined in the following form

$$P_s(q_{QBHS/MRC}) = \frac{2M-1}{M} Q\left(\sqrt{\frac{6}{M^2-1} q_{QBHS/MRC}}\right). \tag{71}$$

Thus, we obtain the closed form expression for the SER of *M*-ary PAM system employing the GR with quadrature subbranch HS/MRC and HS/MRC schemes that agrees with (8.136) and (8.138) in [55]. If $M=2$, the average BER performance of coherent binary phase-shift keying (BPSK) wireless communication system using the quadrature subbranch HS/MRC and HS/MRC schemes under GR implementation can be determined in the following form:

$$P_b = \frac{1}{\pi} \int_0^{0.5\pi} \varphi_{q_{QBHS/MRC}} \left(-\frac{1}{\sin^2\theta}\right) d\theta. \tag{72}$$

3.3.2. MGF of $q_{QBHS/MRC}$

Since all of the 2*N* quadrature branches are i.i.d., the MGF of $q_{QBHS/MRC}$ takes the following form [35]:

$$\varphi_{q_{QBHS/MRC}} = 2L \binom{2N}{2L} \int_0^{\infty} \exp(sq) f(q) [\varphi(s,q)]^{2L-1} [F(q)]^{2(N-L)} dq, \tag{73}$$

where $f(q)$ and $F(q)$ are, respectively, the probability density function (pdf) and the cumulative distribution function (cdf) of *q*, the SNR per symbol, for each quadrature branch, and

$$\varphi(s,q) = \int_q^{\infty} \exp(sx) f(x) dx \tag{74}$$

is the marginal moment generating function (MMGF) of SNR per symbol of a single quadrature branch.

Since g_k and g_{k+N} $(k=1,\ldots,N)$ follow the zero-mean Gaussian distribution with the variance σ_h^2 given by (44), one can show that q_k and q_{k+N} follow the Gamma distribution with pdf given by [49]

$$f(q)=\begin{cases}\frac{1}{\sqrt{q}}\exp\left(-\frac{q}{\bar{q}}\right)\sqrt{\pi\bar{q}}\,, & q\geq 0\\ 0\,, & q\leq 0\,,\end{cases} \tag{75}$$

where

$$\bar{q}=\frac{E_b\sigma_h^2}{\sigma_w^2} \tag{76}$$

is the average SNR per symbol for each diversity branch. The MMGF of SNR per symbol of a single quadrature branch can be determined in the following form:

$$\varphi(s,q)=\frac{1}{\sqrt{1-s\bar{q}}}erfc\left(\sqrt{\frac{1-s\bar{q}}{\bar{q}}q}\right). \tag{77}$$

Moreover, the cdf of q becomes

$$F(q)=1-\varphi(0,q)=1-erfc\left(\sqrt{\frac{q}{\bar{q}}}\right), \tag{78}$$

where $erfc(x)$ is the complimentary error function.

3.4. PCF determination

3.4.1. AWGN channel

In this section, we determine the PCF at the GR output with the first stage of PPIC. From [43], the linear minimum mean-square error (MMSE) solution of PCF for the first stage of PPIC is given by

$$p_{1,\text{opt}}=\frac{\sigma_{2,0}^2-\rho_1\sigma_{1,1}\sigma_{2,0}}{\sigma_{1,1}^2+\sigma_{2,0}^2-2\rho_1\sigma_{1,1}\sigma_{2,0}}, \tag{79}$$

where

http://dx.doi.org/10.5772/58990

$$\sigma_{1,1}^2 = E\{(I_k + \zeta_k - \hat{I}_k)^2\} \tag{80}$$

is the power of residual MAI plus the total noise component forming at the GR output at the first stage;

$$\sigma_{2,0}^2 = E\{(I_k + \zeta_k)^2\} \tag{81}$$

is the power of true MAI plus the total noise component forming at the GR output (also called the 0-th stage);

$$\rho_1 \sigma_{1,1} \sigma_{2,0} = E\{(I_k + \zeta_k - \hat{I}_k)(I_k + \zeta_k)\} \tag{82}$$

is a correlation between these two MAI terms. It can be rewritten as

$$\begin{aligned}
p_{1,\text{opt}} &= \frac{E\{(I_k+\zeta_k)\hat{I}_k\}}{E\{\hat{I}_k^2\}} = \frac{1}{\frac{1}{N}\sum\limits_{u\neq l}^{K} S_u^2 + \sum\limits_{u\neq l}^{K}\sum\limits_{v\neq l,u}^{K} S_u S_v E\{\rho_{ul}\rho_{vl}\hat{b}_u^{(0)}\hat{b}_v^{(0)}\}} \\
&\times\Big\{\frac{1}{N}\sum_{u\neq l}^{K} A_u^2(1-2P_{e,u}) + \sum_{u\neq l}^{K}\sum_{v\neq l,u}^{K} S_u S_v E\{\rho_{ul}\rho_{vl}\hat{b}_u^{(0)}\hat{b}_v^{(0)}\} + \sum_{v\neq l}^{K} S_v E\{\rho_{vl}\zeta_l\hat{b}_v^{(0)}\}\Big\} \\
&= \frac{E\left\{\left[\sum\limits_{j=1,j\neq k}^{K}\alpha_j^2 A_j b_j \rho_{kj} + \int\limits_0^{T_b}\alpha_k^2[w_{k_{AF}}^2(t) - w_{k_{PF}}^2(t)]\,dt\right]\sum\limits_{j=1,j\neq k}^{K}\alpha_j^2 A_j \hat{b}_j^{(0)}\rho_{kj}\right\}}{E\left\{\left[\sum\limits_{j=1,j\neq k}^{K}\alpha_j^2 A_j \hat{b}_j^{(0)}\rho_{kj}\right]^2\right\}} \\
&= \frac{1}{\frac{1}{N}\sum\limits_{i\neq k}^{K}\alpha_i^4 A_i^2 + \sum\limits_{i\neq k}^{K}\sum\limits_{j\neq k,l}^{K}\alpha_i^2 A_i \alpha_j^2 A_j E\{\rho_{ik}\rho_{jk}\hat{b}_i^{(0)}\hat{b}_j^{(0)}\}} \\
&\times\Big\{\frac{1}{N}\sum_{i\neq k}^{K}\alpha_i^4 A_i^2(1-2P_{e,i}) + \sum_{i\neq k}^{K}\sum_{j\neq k,i}^{K}\alpha_i^2 A_i\alpha_j^2 A_j E\{\rho_{ik}\rho_{jk}\hat{b}_i^{(0)}\hat{b}_j^{(0)}\} + \sum_{j\neq k}^{K}\alpha_j^2 A_j E\Big\{\rho_{jk}\hat{b}_j^{(0)}\int_0^{T_b}\alpha_k^2[w_{k_{AF}}^2(t) - w_{k_{PF}}^2(t)]\,dt\Big\}\Big\},
\end{aligned} \tag{83}$$

where $P_{e,i}$ is the BER of user i at the corresponding GR output;

$$E\{\hat{b}_i^{(0)}\hat{b}_i^{(0)}\} = 1 - 2P_{e,i} \quad \text{and} \quad E\{\rho_{ik}^2\} = N^{-1}. \tag{84}$$

The PCF $p_{1,\text{opt}}$ can be regarded as the normalized correlation between the true MAI plus the total noise component forming at the GR output and the estimated MAI. Assume that

$$\mathbf{b} = \{b_k\}_{k=1}^{K} \tag{85}$$

is the data set of all users;

$$\boldsymbol{\rho} = \{\rho_{ik}\}_{i,k=1}^{K} \tag{86}$$

is the correlation coefficient set of random sequences;

$$f_{\tilde{b}_i^{(0)}|\mathbf{b},\boldsymbol{\rho}}(\tilde{b}_i^{(0)} \mid \mathbf{b},\boldsymbol{\rho}) = \mathcal{N}(E\{\tilde{b}_i^{(0)} \mid \mathbf{b},\boldsymbol{\rho}\}, 4\alpha^4\sigma_w^4) \tag{87}$$

is the conditional normal pdf of $\tilde{b}_i^{(0)}$ given $\mathbf{b}$ and $\boldsymbol{\rho}$ and $f_{\tilde{b}_i^{(0)},\tilde{b}_j^{(0)}|\mathbf{b},\boldsymbol{\rho}}(\tilde{b}_i^{(0)},\tilde{b}_j^{(0)} \mid \mathbf{b},\boldsymbol{\rho})$ is the conditional joint normal pdf of $\tilde{b}_i^{(0)}$ and $\tilde{b}_j^{(0)}$ given $\mathbf{b}$ and $\boldsymbol{\rho}$. Following the derivations in [43], the expectation terms with hard decisions in (83) can be evaluated based on Price's theorem [49] as follows

$$E\{\rho_{ik}\rho_{jk}\hat{b}_i^{(0)}\hat{b}_j^{(0)}\} = E\{E\{E\{\rho_{ik}\rho_{jk}\hat{b}_i^{(0)}\hat{b}_j^{(0)} \mid \mathbf{b},\boldsymbol{\rho}\} \mid \boldsymbol{\rho}\}\} = E\{E\{\rho_{ik}\rho_{jk}\hat{b}_i^{(0)}(2Q_j - 1) \mid \boldsymbol{\rho}\}\}; \tag{88}$$

$$E\{\rho_{jk}\zeta_k\hat{b}_j^{(0)}\} = E\{E\{E\{\rho_{jk}\zeta_k\hat{b}_j^{(0)} \mid \mathbf{b},\boldsymbol{\rho}\} \mid \boldsymbol{\rho}\}\} = 4\alpha^4\sigma_w^4 E\{E\{\rho_{jk}^2 f_{\tilde{b}_j^{(0)}|\mathbf{b},\boldsymbol{\rho}}(0 \mid \mathbf{b},\boldsymbol{\rho}) \mid \boldsymbol{\rho}\}\}; \tag{89}$$

$$\begin{aligned} E\{\rho_{ul}\rho_{vl}\hat{b}_u^{(0)}\hat{b}_v^{(0)}\} &= E\{E\{E\{\rho_{ik}\rho_{jk}\hat{b}_i^{(0)}\hat{b}_j^{(0)} \mid \mathbf{b},\boldsymbol{\rho}\} \mid \boldsymbol{\rho}\}\} = E\{E\{\rho_{ik}\rho_{jk}[16\rho_{ij}\alpha^4\sigma_w^4 f_{\tilde{b}_i^{(0)},\tilde{b}_j^{(0)}|\mathbf{b},\boldsymbol{\rho}}(0,0 \mid \mathbf{b},\boldsymbol{\rho}) + \\ &+ (2Q_i - 1)(2Q_j - 1)] \mid \boldsymbol{\rho}\}\}, \end{aligned} \tag{90}$$

where

$$Q_k = Q\Big(-\frac{M\{\tilde{b}_k^{(0)} \mid \mathbf{b},\boldsymbol{\rho}\}}{2\alpha^2\sigma_w^2}\Big) \tag{91}$$

and

$$Var\{\zeta_k\} = 4\alpha_k^4\sigma_w^4 \tag{92}$$

is the total background noise variance forming at the GR output taking into account multipath fading channel; σ_w^2 is the additive Gaussian noise variance forming at the PF and AF outputs of GR linear tract; the Gaussian Q- function is given by (68).

Although numerical integration in [43, 56] can be adopted for determining the optimal PCF $p_{1,\text{opt}}$ for the first stage based on (83)-(90), it requires huge computational complexity. To simplify this problem, we assume that the total background noise forming at the GR output can be considered as a constant factor and may be small enough such that the Q functions in (88) and (90) are all constants and (89) can be approximated to zero. That is

$$4\alpha_k^4\sigma_w^4 << \min_{\{\alpha_k A_k,\boldsymbol{\rho}\}}(E\{\tilde{b}_i^{(0)}\mid \mathbf{b},\boldsymbol{\rho}\})^2 = 4\alpha_m^4 A_m^2 N^{-2}, \tag{93}$$

where [57]

$$\begin{cases} \alpha_m^2 A_m = \min \alpha_k^2 A_k \ ; \\ \sum\limits_{k\neq m}^{K} \alpha_k^2 A_k b_k \rho_{kl} = -\alpha_m^2 A_m b_m \rho'_{mk} \ ; \\ \min|\rho_{mk} - \rho'_{mk}| = \dfrac{2}{N} \ . \end{cases} \tag{94}$$

With this, we can rewrite (88) and (90) as follows:

$$E\{E\{\rho_{ik}\rho_{jk}\hat{b}_i^{(0)}(2Q_j - 1)\mid \boldsymbol{\rho}\}\} = B_1 E\{\rho_{ik}\rho_{jk}\}E\{\hat{b}_i^{(0)}\mid \boldsymbol{\rho}\} = 0; \tag{95}$$

$$\begin{aligned} &E\{E\{\rho_{ik}\rho_{jk}[16\rho_{ik}\alpha^4\sigma_w^4 f_{\tilde{b}_i^{(0)},\tilde{b}_j^{(0)}|\mathbf{b},\boldsymbol{\rho}}(0,0\mid \mathbf{b},\boldsymbol{\rho}) + (2Q_i - 1)(2Q_j - 1)]\mid \boldsymbol{\rho}\}\} \\ &= E\{E\{16\alpha^4\sigma_w^4\rho_{ik}\rho_{jk}\rho_{ij} f_{\tilde{b}_i^{(0)},\tilde{b}_j^{(0)}|\mathbf{b},\boldsymbol{\rho}}(0,0\mid \mathbf{b},\boldsymbol{\rho})\mid \boldsymbol{\rho}\}\} + B_2 E\{E\{\rho_{ik}\rho_{jk}\mid \boldsymbol{\rho}\}\} \\ &= E\{E\{4\alpha^4\sigma_w^4\rho_{ik}\rho_{jk}\rho_{ij} f_{\tilde{b}_i^{(0)},\tilde{b}_j^{(0)}|\mathbf{b},\boldsymbol{\rho}}(0,0\mid \mathbf{b},\boldsymbol{\rho})\mid \boldsymbol{\rho}\}\}, \end{aligned} \tag{96}$$

where B_1 and B_2 are constants. According to assumptions made above, $f_{\tilde{b}_u^{(0)},\tilde{b}_v^{(0)}|\mathbf{b},\boldsymbol{\rho}}(0,0\mid \mathbf{b},\boldsymbol{\rho})$ can be expressed by

$$f_{\tilde{b}_i^{(0)},\tilde{b}_j^{(0)}|\mathbf{b},\boldsymbol{\rho}}(0,0\mid \mathbf{b},\boldsymbol{\rho}) = \frac{\exp(-0.5\boldsymbol{m}_b^T\boldsymbol{B}_b^{-1}\boldsymbol{m}_b)}{8\pi\alpha^4\sigma_w^4\sqrt{1-\rho_{ij}^2}}, \tag{97}$$

where

$$\mathbf{m}_b = [E\{\widetilde{b}_i^{(0)}\mid \mathbf{b},\boldsymbol{\rho}\}, E\{\widetilde{b}_j^{(0)}\mid \mathbf{b},\boldsymbol{\rho}\}]^T \tag{98}$$

and

$$\mathbf{B}_b = E\{(\widetilde{\mathbf{b}} - \mathbf{m}_b)(\widetilde{\mathbf{b}} - \mathbf{m}_b)^T\} \tag{99}$$

with

$$\tilde{\mathbf{b}} = [\tilde{b}_i^{(0)}, \tilde{b}_j^{(0)}]^T. \tag{100}$$

Since $\mathbf{B}_b^{-1}$ is a positive semi-definite matrix, i.e.

$$\boldsymbol{m}_b^T \boldsymbol{B}_b^{-1} \boldsymbol{m}_b \geq 0, \tag{101}$$

we can have

$$0 < f_{\tilde{b}_i^{(0)}, \tilde{b}_j^{(0)} | \mathbf{b}, \boldsymbol{\rho}}(0,0 \,|\, \mathbf{b}, \boldsymbol{\rho}) \leq \max_{\substack{\rho_{ij} \\ \rho_{ij} \neq \pm 1}} \frac{1}{8\pi\alpha^4 \sigma_w^4 \sqrt{1-\rho_{ij}^2}}. \tag{102}$$

With the above results,

$$\min_{\rho_{ij}, \rho_{ij} \neq \pm 1} \sqrt{1-\rho_{ij}^2} = \frac{2\sqrt{N-1}}{N}. \tag{103}$$

where [57]

$$\rho_{ij} = 1 - 2N^{-1} \quad \text{or} \quad -1 + 2N^{-1} \tag{104}$$

and

$$E\{\rho_{ik}\rho_{jk}\rho_{ij}\} = \sum_{m=1}^{N}\sum_{p=1}^{N}\sum_{q=1}^{N} E\{c_{im}c_{km}c_{jp}c_{kp}c_{iq}c_{jq}\} = \sum_{m=1}^{N} N^{-3} = N^{-2}. \tag{105}$$

Thus, we can derive a range of $p_{1,\text{opt}}$ as follows

$$\frac{\sum_{i \neq k}^{K} \alpha_i^4 A_i^2 (1 - 2P_{e,u})}{\sum_{i \neq k}^{K} \alpha_i^4 A_i^2 + \frac{1}{\pi\sqrt{N-1}} \sum_{i \neq k}^{K} \sum_{j \neq l,i}^{K} \alpha_i^2 A_i \alpha_j^2 A_j} \leq p_{1,\text{opt}} < 1 - \frac{2\sum_{i \neq k}^{K} \alpha_i^4 A_i^2 P_{e,u}}{\sum_{i \neq k}^{K} \alpha_i^4 A_i^2}. \tag{106}$$

If the power control is perfect, i.e.

$$\alpha_i^2 A_i = \alpha_j^2 A_j = \alpha^2 A \quad \text{and} \quad P_{e,i} = P_e \tag{107}$$

and P_e is approximated by the BER of high SNR case, i.e., the $Q\left(\sqrt{\frac{N}{K-1}}\right)$ function [58, 59], (83) can be rewritten as

$$\frac{1-2Q\left(\sqrt{\frac{N}{K-1}}\right)}{1+\frac{K-2}{\pi\sqrt{N-1}}} \le p_{1,\mathrm{opt}} < 1-2Q\left(\sqrt{\frac{N}{K-1}}\right). \tag{108}$$

It is interesting to see that the lower and upper boundary values can be explicitly calculated from the processing gain N and the number of users K.

3.4.2. Multipath channel

Based on representation in (37), we can obtain the received signal vector in the following form:

$$\mathbf{x}(t) = \sum_{k=1}^{K} A_k(t) b_k \mathbf{h}_k + \mathbf{w}(t). \tag{109}$$

Introduce the following notation for the correlation coefficient

$$\varpi_{jk} = \boldsymbol{h}_j^T \boldsymbol{h}_k \qquad \text{and} \qquad \varpi_k = \varpi_{kk}. \tag{110}$$

In commercial DS-CDMA wireless communication systems, the users' spreading codes are often modulated with another code having a very long period. As far as the received signal is concerned, the spreading code is not periodic. In other words, there will be many possible spreading codes for each user. If we use the result derived above, we then have to calculate the optimum PCFs for each possible code and the computational complexity will become very high. Since the period of the modulating code is usually very long, we can treat the code chips as independent random variables and approximate the correlation coefficient ϖ_{jk} given by (110) as a Gaussian random variable. In this case, the GR output for the first stage can be presented in the following form:

$$Z_k(t) = A_k(t) b_k \boldsymbol{h}_k^T \boldsymbol{h}_k + \sum_{j=1, j\ne k}^{K} A_j(t) b_j \boldsymbol{h}_j^T \boldsymbol{h}_k + \zeta_k(t) = A_k(t) b_k \varpi_k + \sum_{j=1, j\ne k}^{K} A_j(t) b_j \varpi_k + \zeta_k(t), \tag{111}$$

where the background noise $\zeta_k(t)$ forming at the GR output is given by (56).

Evaluating the GR output process given by (111), based on the well-known results, for example, discussed in [60], we can define the BER performance for the user k in the following form:

$$P_b^{(k)} = 0.5P(Z_k \mid b_k = 1) + 0.5P(Z_k \mid b_k = -1) = P(Z_k \mid b_k = 1). \tag{112}$$

In (112), we assume that the occurrence of probabilities for $b_k = 1$ and $b_k = -1$ are equal, and that the error probabilities for $b_k = 1$ and $b_k = -1$ are also equal. As we can see from (111), there are three terms. The first term corresponds to the desired user bit. If we let $b_k = 1$, it is a deterministic value. The third term in (111) given by (56) corresponds to the GR background noise interference which pdf is defined in [3, Chapter 3, pp. 250–263, 324–328]. The second term in (111) corresponds to the interference from other users and is subjected to the binomial distribution. Note that correlation coefficients in (111) are small and DS-CDMA wireless communication systems are usually operated in low *SNR* environments. The variance of the second term is then much smaller in comparison with the variance of the third term. Thus, we can assume that Z_k conditioned on $b_k = 1$ can be approximated by Gaussian distribution, as shown in [3, Chapter 3, pp. 250–263, 324–328] and [13]. Then the BER performance takes the following form

$$P(Z_k) = Q\left\{\sqrt{\frac{E_{\mathscr{X}}\{\mathscr{M}_k^{(l)}\}}{E_{\mathscr{X}}\{\mathscr{V}_k^{(l)}\}}}\right\}, \tag{113}$$

where $E_{\mathscr{X}}\{\cdot\}$ denotes the expectation operator over the spreading code set $\mathscr{X}$ and $\mathscr{M}_k^{(l)}$ and $\mathscr{V}_k^{(l)}$ are the expected squared mean and variance of Z_k, respectively, given the *l*-th possible code in $\mathscr{X}$. Letting

$$R_k = \sum_{j \neq k} q_j \quad \text{and} \quad \Lambda_k = \sum_{j \neq k} \varpi_{jk}^2, \tag{114}$$

where q_j is defined in (61), considering ϖ_{jk} as a Gaussian random variable, we obtain

$$E_{\mathscr{X}}[\mathscr{M}_k^{(l)}] = A_k^2\{E_{\mathscr{X}}[\varpi_k^{(l)}] - p_k E_{\mathscr{X}}[\Lambda_k^{(l)}]\}^2 = A_k^2\{1 - p_k E_{\mathscr{X}}[\Lambda_k^{(l)}]\}^2 \tag{115}$$

and the mathematical expectation of variance as

$$E_{\mathscr{X}}[\mathscr{V}_k^{(l)}] = 4\sigma_w^4\{E_{\mathscr{X}}[\Omega_{1,k}^{(l)}]p_k^2 - 2E_{\mathscr{X}}[\Omega_{2,k}^{(l)}]p_k + E_{\mathscr{X}}[\Omega_{3,k}^{(l)}]\}. \tag{116}$$

Note that the expectations in (115) and (116) are operated on interfering user bits and noise using the correlation coefficient ϖ_{jk} given by (110). The coefficients of $E_{\mathscr{X}}[\mathscr{V}_k^{(l)}]$ are represented by

$$\Omega_{1,k}^{(l)} = R_k\Big[\sum_{j \neq k} \varpi_{jk}\varpi_j + \sum_{j \neq k}\sum_{m \neq j,k} \varpi_{jm}\varpi_{mk}\Big]^2 + \Big[\sum_{j \neq k} \varpi_{jk}^2\varpi_j + \sum_{j \neq k}\sum_{m \neq j,k} \varpi_{jm}\varpi_{mk}\varpi_{jk}\Big]; \tag{117}$$

$$\Omega_{2,k}^{(l)} = R_k\Big[\sum_{j\neq k}\varpi_{jk}^2\varpi_j + \sum_{j\neq k}\sum_{m\neq j,k}\varpi_{jm}\varpi_{mk}\varpi_{jk}\Big] + \sum_{j\neq k}\varpi_{jk}^2; \tag{118}$$

$$\Omega_{3,k}^{(l)} = R_k\sum_{j\neq k}\varpi_{jk}^2 + \varpi_k. \tag{119}$$

The optimal PCF for the user k can be found as

$$p_{k,\text{opt}} = \arg\max_{p_k}\left\{\frac{E_{\mathcal{X}}[\mathcal{M}_k^{(l)}]}{E_{\mathcal{X}}[\mathcal{V}_k^{(l)}]}\right\} = \left\{p_{k,\text{opt}} : E_{\mathcal{X}}[\mathcal{V}_k^{(l)}]\frac{dE_{\mathcal{X}}[\mathcal{M}_k^{(l)}]}{dp_k} - E_{\mathcal{X}}\left[\mathcal{M}_k^{(l)}\right]\frac{dE_{\mathcal{X}}[\mathcal{V}_k^{(l)}]}{dp_k} = 0\right\}. \tag{120}$$

Substituting (115)–(119) into (120) and simplifying the result, we obtain the following equation

$$p_{k,\text{opt}} = \frac{E_{\mathcal{X}}[\Omega_{2,k}^{(l)}] - E_{\mathcal{X}}[\Omega_{3,k}^{(l)}]E_{\mathcal{X}}[\Lambda_k^{(l)}]}{E_{\mathcal{X}}[\Omega_{1,k}^{(l)}] - E_{\mathcal{X}}[\Omega_{2,k}^{(l)}]E_{\mathcal{X}}[\Lambda_k^{(l)}]}. \tag{121}$$

Unlike that in AWGN channel, the result for the aperiodic code scenario is more difficult to obtain because there are more correlation terms in (114)–(120) to work with. Before evaluation of the expectation terms in (98), we define some function as follows:

$$\alpha_{jk}(m,n) = \alpha_{j,m}\alpha_{k,n}; \tag{122}$$

$$\tau_{jk}(m,n) = \tau_{j,m} - \tau_{k,n}; \tag{123}$$

$$\psi_{jk}(m,n) = \tilde{\mathbf{a}}_{j,m}^T\tilde{\mathbf{a}}_{k,n}. \tag{124}$$

Thus, (122)–(124) define some relative figures between the m-th channel path of the j-th user and the n-th channel path of the k-th user. The notation $\alpha_{jk}(m,n)$ denotes the path gain product, $\tau_{jk}(m,n)$ is the relative path delay, and $\psi_{jk}(m,n)$ is the code correlation with the relative delay $\tau_{jk}(m,n)$. Expanding (122)–(124), we have seven expectation terms to evaluate. For purpose of illustration, we show how to evaluate the first term, $E_{\mathcal{X}}[\varpi_{jk}^2]$ here. By definition, we have ϖ_{jk} as

$$\varpi_{jk} = \mathbf{h}_j^T\mathbf{h}_k = \left\{\sum_{m=1}^{L}\tilde{\mathbf{a}}_{j,m}\alpha_{j,m}\right\}^T\left\{\sum_{n=1}^{L}\tilde{\mathbf{a}}_{k,n}\alpha_{k,n}\right\} = \sum_{m=1}^{L}\sum_{n=1}^{L}\alpha_{j,m}\alpha_{k,n}\tilde{\mathbf{a}}_{j,m}^T\tilde{\mathbf{a}}_{k,n} = \sum_{m=1}^{L}\sum_{n=1}^{L}\alpha_{jk}(m,n)\psi_{jk}(m,n). \tag{125}$$

The expectation of ϖ_{jk} over all possible codes can be presented in the following form:

$$E_{\mathscr{X}}\{\varpi_{jk}^2\} = E\left\{\sum_{m_1=1}^{L}\sum_{n_1=1}^{L}\sum_{m_2=1}^{L}\sum_{n_2}^{L}\alpha_{jk}(m_1,n_1)\psi_{jk}(m_1,n_1)\alpha_{jk}(m_2,n_2)\psi_{jk}(m_2,n_2)\right\}$$
$$= \sum_{m_1=1}^{L}\sum_{n_1=1}^{L}\sum_{m_2=1}^{L}\sum_{n_2}^{L}\alpha_{jk}(m_1,n_1)\alpha_{jk}(m_2,n_2)E\{\psi_{jk}(m_1,n_1)\psi_{jk}(m_2,n_2)\}. \tag{126}$$

Introduce the following function

$$G_{jk}(m_1,n_1,m_2,n_2) = B^2 E[\psi_{jk}(m_1,n_1)\psi_{jk}(m_2,n_2)]. \tag{127}$$

The coefficient B^2 in (127) is only the normalization constant. Since the spreading codes are seen as random, only if $\tau_{jk}(m_1,n_1)$ is equal to $\tau_{jk}(m_2,n_2)$ will $G_{jk}(m_1,n_1,m_2,n_2)$ be nonzero. Consider a specific set of $\{m_1,n_1,m_2,n_2\}$ such that

$$\tau_{jk}(m_1,n_1) = \tau_{jk}(m_2,n_2) = \tau \ , \qquad \tau \geq 0. \tag{128}$$

In this case, we have

$$G_{jk}(m_1,n_1,m_2,n_2) = B^2 \sum_{\nu=0}^{N-\tau-1} E\{a_{j,\nu+\tau}^2 a_{k,\nu}^2\} = N - \tau. \tag{129}$$

At $\tau < 0$, we have the same result except that the sign of τ in (129) is plus. We can conclude that the function $G_{jk}(m_1,n_1,m_2,n_2)$ in (127) can be written in the following form:

$$G_{jk}(m_1,n_1,m_2,n_2) = \begin{cases} N-|\tau| \ , & \text{if} \quad \tau_{jk}(m_1,n_1) = \tau_{jk}(m_2,n_2) = \tau \\ 0 \ , & \text{otherwise} \end{cases}. \tag{130}$$

Using (126), (127), and (129), we can evaluate $E_{\mathscr{X}}[\varpi_{jk}^2]$ in (117)–(119). The formulations from the other six expectation terms can be obtained by mathematical transformation that is not difficult.

We now provide a simple example to show the multipath effect on the optimal PCFs. Introduce the following notations that are satisfied for all *k*:

$$\boldsymbol{\tau}_k = [0,T]^{\mathrm{T}} \ ; \ \mathbf{a}_k = [\beta,\delta]^{\mathrm{T}} \ ; \quad \text{and} \quad \beta^2 + \delta^2 = 1. \tag{131}$$

Using (131) and taking into consideration that in the case of AWGN channel

$$E_{\mathscr{X}}[\Lambda_k^{(l)}] = \frac{K-1}{N}, \tag{132}$$

at given the l-th possible code in $\mathscr{X}$, we can write for the case of the multipath channel

$$E_{\mathscr{X}}[\Lambda_k^{(l)}] = \frac{K-1}{N} + \frac{2(N-T)\beta^2\delta^2(K-1)}{N^2}; \tag{133}$$

$$\begin{aligned} E_{\mathscr{X}}[\Omega_{1,k}^{(l)}] &= E_{\mathscr{X}}\left\{ R_k^{(l)}\left\{ \sum_{j=1,j\neq k}^{K} \rho_{jk}^{(l)} + \sum_{j=1,j\neq k}^{K}\ \sum_{m=1,m\neq j,k}^{K} \rho_{jm}^{(l)}\rho_{mk}^{(l)} \right\}^2 + \sum_{j=1,j\neq k}^{K} (\rho_{jk}^{(l)})^2 + \sum_{j=1,j\neq k}^{K}\ \sum_{m=1,m\neq j,k}^{K} \rho_{jm}^{(l)}\rho_{mk}^{(l)}\rho_{jk}^{(l)} \right\} \\ &+2(N-T)\beta^2\delta^2\{R_k N^{-4}[N^2+10N+4(N-T)\beta^2\delta^2+2(K-2)(4N+3K+(N-T)\beta^2\delta^2+1] \\ &+(K-1)N^{-3}(N+3K-2)\}+4(N-2T)\beta^4\delta^4[R_k K N^{-4}+6K-12]+R_k N^{-4}(6N-10T)\beta^4\delta^4; \end{aligned} \tag{134}$$

$$\begin{aligned} E_{\mathscr{X}}[\Omega_{2,k}^{(l)}] &= E_{\mathscr{X}}\left\{ R_k^{(l)}\left\{ \sum_{j=1,j\neq k}^{K} (\rho_{jk}^{(l)})^2 + \sum_{j=1,j\neq k}^{K}\ \sum_{m=1,m\neq j,k}^{K} \rho_{jm}^{(l)}\rho_{mk}^{(l)}\rho_{jk}^{(l)} \right\} + \sum_{j=1,j\neq k}^{K} (\rho_{jk}^{(l)})^2 \right\} \\ &+2(N-T)\beta^2\delta^2[R_k N^{-3}(N+3K-2)+(k-1)N^{-2}]; \end{aligned} \tag{135}$$

$$E_{\mathscr{X}}[\Omega_{3,k}^{(l)}] = E_{\mathscr{X}}\left\{ R_k \sum_{j=1,j\neq k}^{K} (\rho_{jk}^{(l)})^2 + 1 \right\} + 2(N-T)\beta^2\delta^2 R_k N^{-2}. \tag{136}$$

Note that the first terms in (133)–(136) correspond to the optimal PCFs in AWGN channel. Other terms are due to the multipath effect. It is evident to see that if $\delta = 0$ we have the case of AWGN channel.

3.5. Simulation results

3.5.1. Selection/maximal-ratio combining

In this section, we discuss some examples of GR performance with quadrature subbranch HS/MRC and HS/MRC schemes and compare with the conventional HS/MRC receiver. The average SER of coherent BPSK and 8-PAM signals under processing by the GR with the quadrature subbranch HS/MRC and HS/MRC schemes as a function of average SNR per symbol per diversity branch for various values of $2L$ and $2N = 8$ is presented in Fig.3. It is seen that the GR SER performance with quadrature subbranch HS/MRC and HS/MRC schemes at $(L,N) = (3,4)$ achieves virtually the same performance as the GR with traditional MRC, and that the performance at $(L,N) = (2,4)$ is typically less than 0.5 dB in SNR poorer than that of GR with traditional MRC in [42]. Additionally, a comparison with the traditional HS/MRC receiver in [29, 30] is made. Advantage of GR implementation in DS-CDMA wireless communication systems is evident.

The average SER of coherent BPSK and 8-PAM signals under processing by GR with quadrature subbranch HS/MRC and HS/MRC schemes as a function of average SNR per symbol per diversity branch for various values of $2N$ at $2L = 4$ is shown in Fig.4. We note the substantial benefits of increasing the number of diversity branches N for fixed L. Comparison with the traditional HS/MRC receiver is made. Advantage of GR implementation in DS-CDMA wireless communication systems is evident.

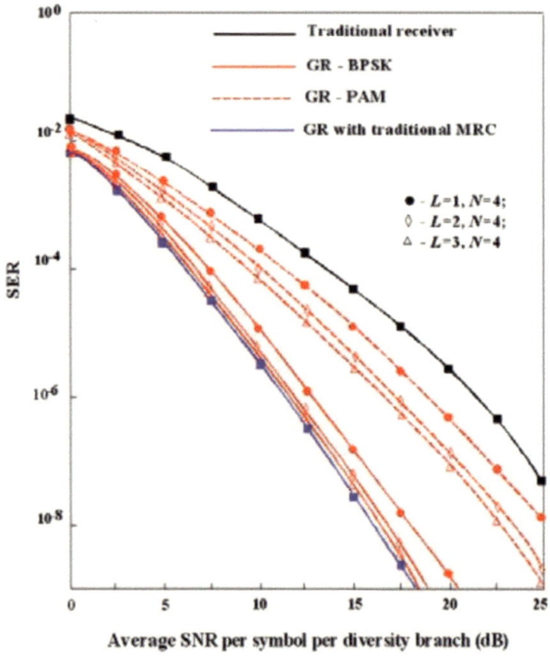

Figure 3. Average SER of coherent BPSK and 8-PAM for GR with quadrature subbranch HS/MRC and HS/MRC schemes versus the average SNR per symbol per diversity for various values of $2L$ with $2N = 8$.

Comparative analysis of average BER as a function of the average SNR per bit per diversity branch of coherent BPSK signals employing GR with quadrature subbranch HS/MRC and HS/MRC schemes and GR with traditional HS/MRC scheme for various values of L with $N = 8$ is presented in Fig. 5. To achieve the same value of average SNR per bit per diversity branch, we should choose $2L$ quadrature branches for the GR with quadrature subbranch HS/MRC, HS/MRC schemes, and L diversity branches for the GR with traditional HS/MRC scheme. Figure 5 demonstrates that the GR BER performance with quadrature subbranch HS/MRC and HS /MRC schemes is much better than that of the GR with traditional HS/MRC scheme, about 0.5 dB to 1.2 dB, when L is less than one half N. This difference decreases with increasing L. This is expected because when $L = N$ we obtain the same performance. Some discussion of increases in GR complexity and power consumption is in order. We first note that GR with quadrature subbranch HS/MRC and HS/MRC schemes requires the same number of antennas as GR with traditional HS/MRC scheme. On the other hand, the former requires twice as many comparators as the latter, to select the best signals for further processing. However, GR designs that process the quadrature signal components will require $2L$ receiver chains for either the GR with quadrature subbranch HS/MRC and HS/MRC schemes or the GR with traditional HS/MRC scheme. Such receiver designs will use only a little additional power, as GR chains consume much more power than the comparators.

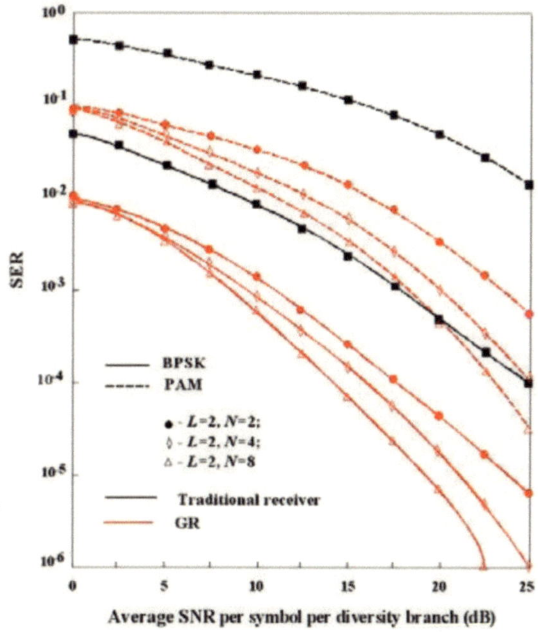

Figure 4. Average SER of coherent BPSK and 8-PAM for GR with quadrature subbranch HS/MRC and HS/MRC schemes versus the average SNR per symbol per diversity for various values of $2N$ with $2L = 4$.

On the other hand, GR designs that implement co-phasing of branch signals without splitting the branch signals into the quadrature components will require L receiver chains for GR with traditional HS/MRC scheme and $2L$ receiver chains for GR with quadrature subbranch HS/MRC and HS/MRC schemes, with corresponding hardware and power consumption increases.

3.5.2. Synchronous DS-CDMA wireless communication system

To demonstrate a usefulness of the optimal PCF range given by (108), we performed a number of simulations for asynchronous DS-CDMA wireless communication system with perfect power control. In simulations, the random spreading codes with length $N = 64$ were used for each user and the number of users was $K = 40$ [61]. Figure 6 shows the BER performance of single-stage hard-decision GR based on PPIC for different magnitudes of SNR and various values of PCF where the optimal PCF for the first stage lies between 0.3169 (lower boundary) and 0.7998 (upper boundary). It can be seen that, for all the SNR cases, the GR based on PPIC using the average of the lower and upper boundary values, i.e., 0.5584, as the PCF, has the close BER performance to that using the optimal PCF. Additionally, comparison of GR implementation in DS-CDMA systems with the conventional detector discussed in [43] is presented.

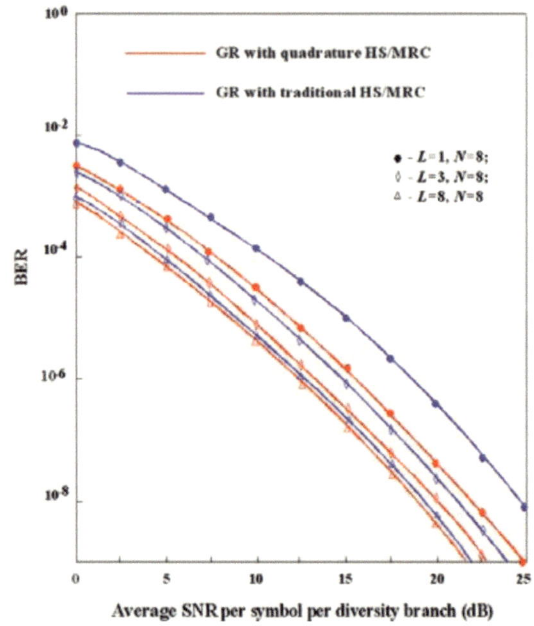

Figure 5. Comparison of the average BER of coherent BPSK and 8-PAM for GR with quadrature subbranch HS/MRC and HS/MRC schemes versus the average SNR per symbol per diversity for various values of $2L$ with $N = 8$.

These results demonstrate us a great superiority of the GR employment over the conventional detector in [43].

Figure 7 shows the BER performance at each stage for the three-stage GR based on the PPIC using different PCFs at the first stage, i.e., the average value and an arbitrary value. PCFs for these two three-stage cases are

$$(a_1, a_2, a_3) = (0.5584, 0.8, 0.9) \quad \text{and} \quad (0.7, 0.8, 0.9) \tag{137}$$

respectively. The results demonstrate that the BER performances of GR employed by DS-CDMA systems for the cases using the proposed PCF at the first stage outperform ones of GR implemented in DS-CDMA system using arbitrary PCF at the first stage. Furthermore, the GR BER performance at the second stage for the case using the proposed PCF at the first stage achieves the GR BER performance of the GR comparable to that of the three-stage GR based on PPIC using an arbitrary PCF at the first stage. Comparison between the AWCN and multipath channels is also presented in Fig.7. We see that in the case of multipath channel, the BER performance is deteriorated. This fact can be explained by the additional correlation terms in (133)–(136).

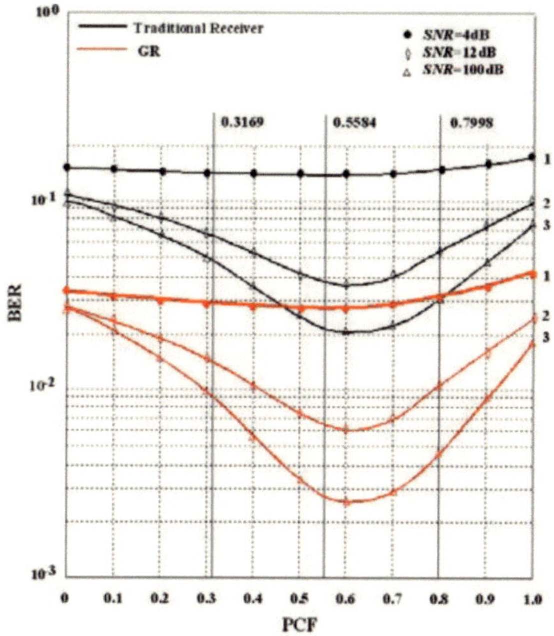

Figure 6. The BER performance of the single-state GR based on PPIC with hard decisions for different SNRs and PCFs.

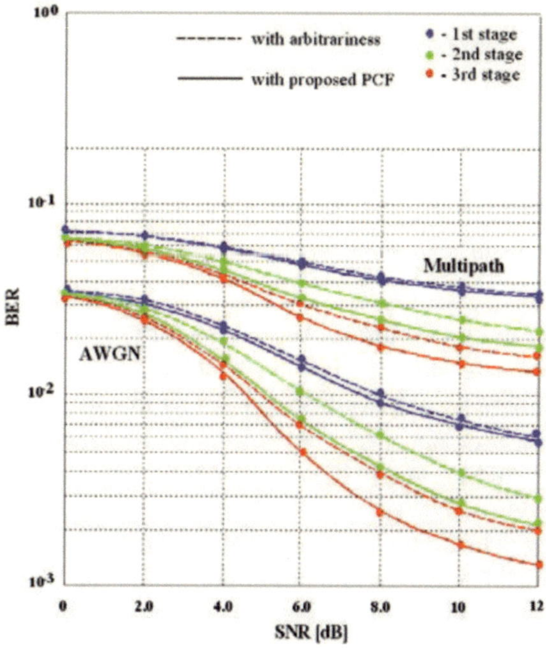

Figure 7. The BER performance at each stage for three-stage GR based on the PPIC with hard decisions for different PCFs at the first stage, i.e., the average value and an arbitrary value: AWGN and multipath channels.

Figure 8 demonstrates the optimal PCF versus the number of users both for the synchronous AWGN and for the multipath channels. We carry out simulation for the AWGN channel under the following conditions: the Gold codes, *SNR*=12 dB, the spreading codes are the periodic and perfect power control. The multipath channel assumed is a two-ray channel with the transfer function

$$W_k(Z) = 0.762 + 0.648Z^{-2} \tag{138}$$

for all users. In the case of multipath channel, we employ aperiodic codes, *SNR* =12 dB, and perfect power control.

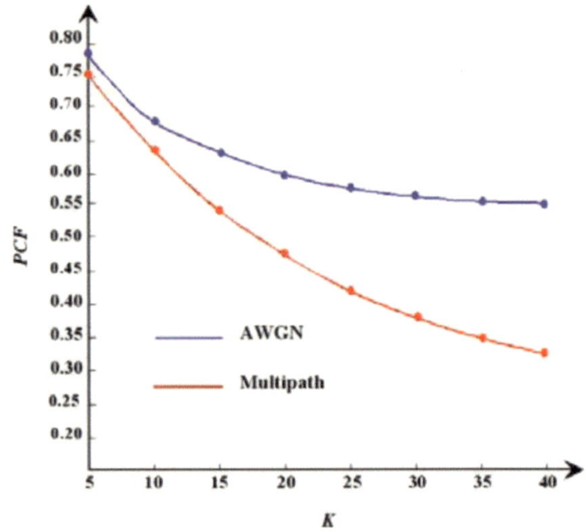

Figure 8. Optimal PCF versus the number of users: the AWGN and multipath channels.

3.6. Conclusions

The GR performance with quadrature subbranch HS/MRC and HS/MRC schemes for a 1-D signal modulation in Rayleigh fading was investigated. The SER of *M*-ary PAM, including coherent BPSK modulation, was derived. Results show the GR with quadrature subbranch HS/MRC and HS/MRC schemes performs substantially better the GR with traditional HS/MRC scheme, particularly, when *L* is smaller than one half *N*, and much better the traditional HS/MRC receiver. We have also derived the optimal PCF range for GR first stage based on the PPIC, which is employed by DS-CDMA system, with hard decisions in multipath fading channel. Computer simulation shows that the BER performance of the GR employed by DS-CDMA wireless communication system with multipath fading channel in the case of periodic code scenario and using the average of the lower and upper boundary values is close to that of the GR of the case using the real optimal PCF, whether the SNR is high or low. It has also been shown that GR employment in DS-CDMA system with multipath fading channel in the case of periodic code scenario allows us to observe a great

http://dx.doi.org/10.5772/58990

superiority over the conventional receiver discussed in [43]. The procedure discussed in [43] is also acceptable for GR employment by DS-CDMA systems. It has also been demonstrated that the two-stage GR based on PPIC using the proposed PCF at the first stage achieves such BER performance comparable to that of the three-stage GR based on PPIC using an arbitrary PCF at the first stage. This means that at the same BER performance, the number of stages (or complexity) required for the multistage GR based on PPIC could be reduced when the proposed PCF is used at the first stage. It can be shown that the proposed PCF selection approach is applicable to multipath fading cases at GR employment in DS-CDMA systems even if no perfect power control is assumed but this is a subject of future work. We have also compared the BER performance at the optimal PCF in the case of AWGN and multipath channels and presented a sensitivity of the BER performance to the values of PCF for both cases.

4. Signal processing with frequency-selective channels

4.1. Brief review

In this section, we consider and study the GR in DS-CDMA wireless communication system with frequency-selective channels. We discuss the linear equalization with the finite impulse response (FIR) beamforming filters and channel estimation and spatially correlation.

4.1.1. Linear equalization and FIR beamforming filters

The use of multiple antennas in wireless communication system attracts significant interest and attention of researchers. Transmit beamforming has received considerable attention because of its simplicity and ability to exploit the benefits of multiple transmit antenna [62]. Information about channel state at the transmitter is generally required for beamforming. At the present time, an impact of noisy and/or quantized information about channel state is a pivot of recent research owing to the fact that the perfect channel state information may not be available at the transmitter [63–65]. In [66], beamforming techniques for DS-CDMA wireless communication systems used a multicarrier approach to cope with frequency-selective fading were also proposed. We should note, however, that the multicarrier techniques are not used in single carrier systems. In this section, we investigate the transmit beamforming for single carrier transmission over frequency-selective fading channels with the perfect channel state information at the GR. A necessity of equalization at the GR is generated by the intersymbol interference (ISI) caused by the channel frequency selectivity. It must be emphasized that the optimum beamforming depends on the equalizer used. As it is well-known, the linear equalization possesses a low complexity. By this reason, we adopt the linear equalization. In comparison and in contrast to [67], we consider the more realistic case of FIR beamforming filters at the GR. Unlike the infinite impulse response case, a closed-form solution for the FIR beamforming filters at the GR does not seem to exist. Because of this, we need to calculate the optimum FIR beamforming filters. Numerical results show that for typical Global System for Mobile Communications/Enhanced Data Rates for GSM Evolution (GSM/EDGE) channels the performance of short GR FIR beamforming filters can be closely approached with infinite impulse response beamforming at the GR and significant gains over single antenna transmission can be achieved.

4.1.2. Channel estimation and spatial correlation

Under consideration of channel estimation and correlation part, we investigate the minimum mean square error (MMSE) GR. It takes an error of maximum likelihood (ML) multiple-input multiple-output (MIMO) channel estimation and GR spatially correlation into consideration in the computation of MMSE GR and log-likelihood ratio (LLR) of each coded bit. Our GR analysis and investigation are based on the following statements:

- It is well-known [68–70] that the existing soft-output MMSE vertical Bell Lab Space Time (V-BLAST) detectors have been designed based on perfect channel estimation. Unfortunately, the estimated MIMO channel coefficient matrix is noisy and imperfect in practical application environment, [71,72]. Therefore, these soft-output MMSE V-BLAST detectors will suffer from performance degradation under practical channel estimation.
- The ML symbol detection scheme is investigated in [72]. It takes into consideration the channel estimation error under condition that the MIMO channel estimation is imperfect. MMSE based on V-BLAST symbol detection algorithm addressing the impact of the channel estimation error is discussed in [73]. However, the channel coding, decision error propagation compensation, and spatially channel correlation are not discussed and considered.

In the present section, we derive the MMSE GR for detecting each transmitted symbol and provide a method to compute the LLR of each coded bit. When compared with the detection scheme discussed in [73], our simulation results show that the MMSE GR can obtain sizable performance gain.

4.2. Problem statement and system model

We consider a MIMO DS-CDMA wireless communication system with N_T transmit and N_R receive antennas. The modulated symbols $b[k]$ are taken from a scalar symbol alphabet $\mathcal{F}$ such as quadriphase-shift keying (QPSK) or quadrature amplitude modulation (QAM), and have the following variance

$$\sigma_b^2 = E\{|b[k]|^2\} = 1, \tag{139}$$

where $E\{\cdot\}$ denotes the mathematical expectation. The coefficients of the FIR beamforming filters of length L_g at the transmit antenna n_t, $1 \le n_t \le N_T$ are denoted by $g_{n_t}[k]$, where $0 \le k \le L_g - 1$ and their energy is normalized to

$$\sum_{n_t=1}^{N_T}\sum_{k=0}^{L_g-1}|g_{n_t}[k]|^2=1. \tag{140}$$

The signal transmitted over antenna n_t at time k is given by

$$a_{n_t}[k]=g_{n_t}[k]\otimes b[k], \tag{141}$$

where $\otimes$ denotes a linear discrete-time convolution. The discrete-time received signal at the receive antenna n_r, $1\le n_r\le N_R$ can be modeled in the following manner

$$s_{n_r}[k]=\sum_{n_t=1}^{N_T}h_{n_t n_r}[k]\otimes a_{n_t}[k]+w_{n_r}[k], \tag{142}$$

where $w_{n_r}[k]$ denotes the spatially and temporally AWGN with zero mean and variance given by

$$\sigma_w^2=E\{|w_{n_r}[k]|^2\}=0.5N_0, \tag{143}$$

where $0.5N_0$ denotes the two-sided power spectral density of the underlying continuous in time passband noise process.

The notation $h_{n_t n_r}[k]$, where $0\le k\le L-1$, represents the overall channel impulse response between the transmit antenna n_t and the receive antenna n_r of length L. In our model, $h_{n_t n_r}[k]$ contains the combined effects of transmit pulse shaping, wireless channel, receive filtering, and sampling. We assume an existence of block fading model, i.e., the channel is constant for the duration of at least one data burst before it changes independently to a new realization. In general, $h_{n_t n_r}[k]$ are spatially and temporally correlated because of insufficient antenna spacing and transmit/receive filtering, respectively. Substituting (141) into (142), we obtain

$$s_{n_r}[k]=h_{n_r}^{eq}[k]\otimes b[k]+w_{n_r}[k], \tag{144}$$

where the equivalent channel impulse response $h_{n_r}^{eq}[k]$ corresponding to the receiving antenna n_r is defined as

$$h_{n_r}^{eq}[k]=\sum_{n_t=1}^{N_T}h_{n_t n_r}[k]\otimes g_{n_t}[k] \tag{145}$$

and has the length $L_{eq} = L + L_g - 1$. Equation (144) shows that the MIMO DS-CDMA wireless communication system with beamforming can be modeled as an equivalent single-input multiple-output (SIMO) system. Therefore, the GR can use the same equalization, channel estimation, and channel tracking techniques as for a single antenna transmission. We assume that the GR employs receive diversity zero forcing or MMSE linear equalization [74].

Let us rewrite the main statements and definitions mentioned above in the matrix form for our convenience in subsequent analysis of channel estimation. Thus, the received signal can be expressed in the following form:

$$\mathbf{s} = \mathbf{Ha} + \mathbf{w} = \sum_{k=1}^{N_T} \mathbf{h}_k a_k + \mathbf{w}, \tag{146}$$

where $\mathbf{s} = [s_1, s_2, \ldots, s_{N_R}]^T$ is the received signal vector; $\mathbf{H} = [\mathbf{h}_1, \mathbf{h}_2, \ldots, \mathbf{h}_{N_T}]^T$ is the $N_R \times N_T$ MIMO channel coefficient matrix with elements $h_{n_t n_r}[k]$ denoting the channel fading coefficient between the n_t -th transmit antenna and the n_r -th receive antenna.

We adopt the following GR spatially correlated MIMO channel model

$$\mathbf{H} = \sqrt{\mathbf{R_r}}\,\mathbf{H}_w \tag{147}$$

with $\mathbf{H}_w$ denoting an independent and identically distributed (i.i.d.) matrix with entries obeying the Gaussian law with zero mean and unit variance, and $\mathbf{R_r}$ is the $N_R \times N_R$ receive array correlation matrix determined by

$$\mathbf{R_r} = \sqrt{\mathbf{R_r}}\left(\sqrt{\mathbf{R_r}}\right)^H. \tag{148}$$

Then, we have

$$E\{\mathbf{HH}^H\} = N_T \mathbf{R_r}. \tag{149}$$

The channel is considered to be flat fading with coherence time of $(N_P + N_D)$ MIMO vector symbols, where N_P symbol intervals are dedicated to pilot matrix $\mathbf{S}_p$ and the remaining N_D to data transmission, where $\mathbf{a} = [a_1, a_2, \ldots, a_{N_T}]^T$ is the transmitted complex signal vector whose element given by (141) is taken from the complex modulation constellation $\mathcal{F}$, because the modulated symbols $b[k]$ are taken from the scalar symbol alphabet $\mathcal{F}$, such as QPSK signal, by mapping the coefficient of FIR beamforming filters $g_{n_t}[k]$ like the coded bit vector

$$\mathbf{g}_{n_t} = \left[\mathbf{g}_{n_t}^1, \mathbf{g}_{n_t}^2, \ldots, \mathbf{g}_{n_t}^{\log_2 M}\right]^T \tag{150}$$

to one modulation symbol belonging to $\mathcal{F}$, i.e., $a_{n_t} = \mathrm{map}(\mathbf{g}_{n_t}) \in \mathcal{F}$.

Meanwhile, we assume that each transmitted symbol is independently taken from the same modulation constellation $\mathcal{F}$ and has the same average energy, i.e.,

$$E\{\mathbf{a}\mathbf{a}^H\} = E_b \mathbf{I}_{N_T}. \tag{151}$$

Finally, $\mathbf{w} = [w_1, w_2, \ldots, n_{N_R}]^T$ is the AWGN vector with covariance matrix determined by

$$\mathbf{K_w} = E\{\mathbf{w}\mathbf{w}^H\} = \sigma_w^2 \mathbf{I}_{N_R}. \tag{152}$$

$\mathbf{I}_{N_T}$ and $\mathbf{I}_{N_R}$ are the identity matrices.

4.3. FIR beamforming for GR with linear equalization

According to [75], the unbiased *SNR* for linear equalization with the optimum infinite impulse response equalizer filters at the GR back end is given by

$$SNR(\mathbf{g}) = \frac{\sigma_b^4}{\sigma_e^4} - \chi, \tag{153}$$

where σ_b^2 is given by (139) and the linear equalization error variance σ_e^2 will be defined below. We note that the assumption of infinite impulse response linear equalization filters at the GR back end is not a major restriction, since typically FIR linear equalization filters of length equal to $L_F = 4L_{eq}$ can approach closely the performance of infinite impulse response linear equalization filters. In (153) we consider the constant $\chi = 0$ for the case of zero forcing linear equalization and $\chi = 1$ for the case of MMSE linear equalization, respectively [74]. In (153) the beamforming filter vector

$$\mathbf{g} = [g_1(0) g_1(1) \cdots g_1(L_g - 1) g_2(0) \cdots g_{N_T}(L_g - 1)]^T \tag{154}$$

consists of the coefficients of all beamforming filters.

The GR linear equalization error variance defined based on results discussed in [76] is given by

$$\sigma_e^4 = 4\sigma_w^4 \int_{-0.5}^{0.5} \frac{1}{\sum_{n_r=1}^{N_R} |H_{n_r}^{eq}(f)|^2 + \mu} df, \tag{155}$$

where $\mu = 0$ and $\mu = 4\sigma_w^4 / \sigma_b^4$ are valid for the case of zero-forcing linear equalization and for the case of MMSE linear equalization, respectively. Furthermore, the frequency response $H_{n_r}^{eq}(f) = \mathcal{F}\{h_{n_r}^{eq}(k)\}$ of the equivalent channel can be defined in the following form

$$H_{n_r}^{eq}(f) = \mathbf{q}^H(f)\mathbf{H}_{n_r}\mathbf{g}, \tag{156}$$

where the subscript H means the Hermitian transpose,

$$\mathbf{q}(f) = \{1 \ \exp(j2\pi f) \cdots \exp[j2\pi f(L_{eq} - 1)]\}^T, \tag{157}$$

$$\mathbf{H}_{n_r} = [\mathbf{H}_{1n_r} \mathbf{H}_{2n_r} \cdots \mathbf{H}_{N_T n_r}]^T \tag{158}$$

and $\mathbf{H}_{n_t n_r}$ is a $L_{eq} \times L_g$ column-circulant matrix with the vector $[h_{n_t n_r}(0) \ldots h_{n_t n_r}(L-1)\mathbf{0}_{L_g-1}^T]^T$ in the first column. Therefore, the GR *SNR* with the zero-forcing linear equalization and MMSE linear equalization can be presented in the following form:

$$SNR(\mathbf{g}) = \frac{\sigma_b^4}{4\sigma_w^4 \int_{-0.5}^{0.5} \frac{1}{\mathbf{g}^H \mathbf{G}(f)\mathbf{g} + \zeta} df} - \chi \tag{159}$$

with the $N_T L_g \times N_T L_g$ matrix

$$\mathbf{G}(f) = \sum_{n_r=1}^{N_R} \mathbf{H}_{n_r}^H \mathbf{d}(f)\mathbf{d}^H(f)\mathbf{H}_{n_r}. \tag{160}$$

The optimum beamforming filter vector $\mathbf{g}_{opt}$ shall maximize $SNR(\mathbf{g})$ subject to power constraint $\mathbf{g}^H\mathbf{g} = 1$. Unfortunately, this optimization problem is not convex, i.e. the standard tools from convex optimization cannot be applied. Nevertheless, the Lagrangian of the optimization problem can be formulated in the following form:

$$L(\mathbf{g}) = SNR(\mathbf{g}) + \mu\mathbf{g}^H\mathbf{g}, \tag{161}$$

where μ denotes the Lagrange multiplier. The optimum vector $\mathbf{g}_{opt}$ has to satisfy the following equality

$$\frac{\partial L(\mathbf{g})}{\partial \mathbf{g}^*} = \mathbf{0}_{N_T L_g}, \tag{162}$$

which leads to the nonlinear eigenvalue problem, namely,

$$\left[\int_{-0.5}^{0.5} \frac{\mathbf{G}(f)}{[\mathbf{g}_{opt}^H \mathbf{G}(f)\mathbf{g}_{opt} + \zeta]^2} df\right] \mathbf{g}_{opt} = \tilde{\mu}\mathbf{g}_{opt}, \tag{163}$$

with the eigenvalue $\tilde{\mu}$. Unfortunately, (163) does not seem to have a closed-form solution.

Therefore, we use the following gradient algorithm for calculation of the optimum FIR beamforming filters at the GR, which recursively improves an initial beamforming filter vector $\mathbf{g}_0$. The main statements of the gradient algorithm are:

1. Let $i = 0$ and initialized the beamforming filter vector with a suitable $\mathbf{g}_0$ satisfying $\mathbf{g}_0^H \mathbf{g}_0 = 1$.
2. Update the beamforming filter vector

$$\tilde{\mathbf{g}}_{i+1} = \mathbf{g}_i + \delta\left[\int_{-0.5}^{0.5} \frac{\mathbf{G}(f)}{[\mathbf{g}_i^H \mathbf{G}(f)\mathbf{g}_i + \zeta]^2} df\right] \mathbf{g}_i, \tag{164}$$

where δ_i is a suitable adaptation step size.

3. Normalize the beamforming filter vector

$$\mathbf{g}_{i+1} = \frac{\tilde{\mathbf{g}}_{i+1}}{\sqrt{\tilde{\mathbf{g}}_{i+1}^H \tilde{\mathbf{g}}_{i+1}}}. \tag{165}$$

4. If $1 - |\mathbf{g}_{i+1}^H \mathbf{g}_i| < \varepsilon$, go to Step 5, otherwise increment $i \to i+1$ and go to Step 2.
5. $\mathbf{g}_{i+1}$ are the desired beamforming filter vector.

For the termination constant ε in Step 4 a suitably small value should be chosen, e.g. $\varepsilon = 10^{-4}$. Ideally the adaptation step size δ_i should be optimized to maximize the speed of convergence. Here, we follow [77] and choose δ_i proportional to λ_i^{-1}, where λ_i is the maximal eigenvalue of the matrix

$$\left[\int_{-0.5}^{0.5} \frac{\mathbf{G}(f)}{[\mathbf{g}_i^H \mathbf{G}(f)\mathbf{g}_i + \zeta]^2} df\right] \tag{166}$$

in iteration *i*. In particular, we found empirically that $\delta_i = 0.01\lambda_i^{-1}$ is a good choice.

Because of non-convexity of the underlying optimization problem, we cannot guarantee that the gradient algorithm converges to the global maximum. However, adopting the initialization procedure explained below, the solution found by this gradient algorithm seems to be close to optimum, i.e., if L_g is chosen sufficiently large the FIR beamforming filters obtained with the gradient algorithm approach and the performance of the optimal infinite impulse response beamforming filters at the GR was discussed in [52].

We found empirically that a convergence to the optimum or a close to optimum solution is achieved if the beamforming filter length is gradually increased. If the desired beamforming filter length is L_g, the gradient algorithm is executed L_g times. The beamforming filter vector is initialized with the normalized all-ones vector of size N_T for the first execution $(\upsilon = 1)$ of the gradient algorithm. For the υ-th execution, $2 \leq \upsilon \leq L_g$, the first $(\upsilon - 1)$ beamforming filters coefficients of each antenna are initialized with the optimum beamforming filter coefficients for that antenna found in the $(\upsilon - 1)$-th execution of the gradient algorithm and the υ-th coefficients are initialized with zero. In each execution step υ, the algorithm requires typically less than 50 iterations to converge, i.e., the overall complexity of the algorithm are on the order of 50 L_g.

4.4. MMSE GR

4.4.1. Channel estimation

It was proved that for ML MIMO channel estimator the optimal pilot matrix minimizing the mean square estimation error is an orthogonal pilot matrix [71, 72]. Under the use of the pilot matrix, i.e.,

$$\mathbf{S}_p \mathbf{S}_p^H = E_P N_P \mathbf{I}_{N_T}, \tag{167}$$

where $N_P \geq N_T$ and E_P is the energy of each pilot symbol, the estimated channel matrix can be expressed as [71, 72] $\hat{\mathbf{H}} = \mathbf{H} + \Delta\mathbf{H}$, where

$$\Delta\mathbf{H} = \mathbf{w}\mathbf{S}_p^H (E_P N_P)^{-1} \tag{168}$$

is the channel estimation error matrix, which is correlated with the matrix $\mathbf{H}$ and with entries subjected to Gaussian distribution with zero mean and variance

$$\sigma_{\Delta h}^2 = \sigma_w^2 (E_P N_P)^{-1}, \tag{169}$$

which is determined independently of instantaneous channel realization. We can conclude that $\hat{\mathbf{H}}$ is a complex Gaussian matrix with zero mean and covariance matrix

$$Cov[\hat{\mathbf{H}}] = E\{\hat{\mathbf{H}}\hat{\mathbf{H}}^H\} = N_T(\mathbf{R}_\mathbf{r} + \sigma_{\Delta h}^2 \mathbf{I}_{N_R}). \tag{170}$$

Let $\mathbf{h}_m, \Delta\mathbf{h}_m$ and $\hat{\mathbf{h}}_m, (m = 1,2,\ldots,N_T)$ denote the m-th column of matrices $\mathbf{H}$, $\Delta\mathbf{H}$ and $\hat{\mathbf{H}}$, respectively. Then, by the important properties of complex Gaussian random vector [78] and with some manipulations, we obtain

$$E\{\Delta\mathbf{h}_m \mid \hat{\mathbf{h}}_m\} = \sigma_{\Delta h}^2 \hat{\mathbf{h}}_m (\mathbf{R}_\mathbf{r} + \sigma_{\Delta h}^2 \mathbf{I}_{N_R})^{-1}; \tag{171}$$

$$Cov[\Delta\mathbf{h}_m \Delta\mathbf{h}_m^H \mid \hat{\mathbf{h}}_m] = \sigma_{\Delta h}^2 \mathbf{I}_{N_R} - \frac{\sigma_{\Delta h}^4}{\mathbf{R}_\mathbf{r} + \sigma_{\Delta h}^2 \mathbf{I}_{N_R}}. \tag{172}$$

4.4.2. Computation of MMSE GR

Let $k_i \in \{1,2,\ldots,N_T\}$ be the index of i-th detected spatial data stream according to the maximal post-detection *SINR* ordering rule. Denote μ_{a_j} and $\sigma_{a_j}^2$ as the mean and variance of the signal a_j, respectively, which can be determined by *a posteriori* symbol probability estimation as in [70]. By performing the soft interference cancellation (SIC) [70] and considering channel estimation error, the corresponding interference-cancelled received signal vector $\tilde{\mathbf{s}}_{k_i}$ can be determined in the following form:

$$\tilde{\mathbf{s}}_{k_i} = \mathbf{H}\mathbf{a} - \sum_{j=k_1}^{k_{i-1}} \hat{\mathbf{h}}_j \mu_{a_j} + \mathbf{w}_{PF} = \sum_{j=k_i}^{k_{N_T}} (\hat{\mathbf{h}}_j - \Delta\mathbf{h}_j) a_j + \sum_{j=k_1}^{k_{i-1}} \hat{\mathbf{h}}_j (a_j - \mu_{a_j}) - \sum_{j=k_1}^{k_{i-1}} \Delta\mathbf{h}_j a_j + \mathbf{w}_{PF}, \tag{173}$$

where $\mathbf{w}_{PF}$ is the noise forming at the PF output of GR front end linear system.

Then, conditionally on $\hat{\mathbf{H}}$, the MMSE GR output is given as [3, 51]

$$\tilde{\mathbf{Y}}_i = \frac{E\{2a_{k_i}\tilde{\mathbf{s}}_{k_i}^H \mid \hat{\mathbf{H}}\} - E\{\tilde{\mathbf{s}}_{k_i}\tilde{\mathbf{s}}_{k_i}^H \mid \hat{\mathbf{H}}\} + E\{\mathbf{w}_{AF_i}\mathbf{w}_{AF_i}^H\}}{E\{\tilde{\mathbf{s}}_{k_i}\tilde{\mathbf{s}}_{k_i}^H \mid \hat{\mathbf{H}}\}}, \tag{174}$$

where $\mathbf{w}_{AF}$ is the reference zero mean Gaussian noise with *a priori* information **a "no" signal** and with the following covariance matrix in a general case [1, 3]

$$E\{\mathbf{w}_{PF}\mathbf{w}_{PF}^H\} = E\{\mathbf{w}_{AF}\mathbf{w}_{AF}^H\} = \sigma_w^2 \mathbf{I}_{N_R}, \tag{175}$$

because the AF and PF of GR front end linear system do not change the statistical parameters of input process (Gaussian noise, for example). Thus, according to (173) and (175), we can write

$$E\{\tilde{\mathbf{s}}_{k_i}\tilde{\mathbf{s}}_{k_i}^H \mid \hat{\mathbf{H}}\} = \Big\{\sum_{j=k_i}^{k_{N_T}}[\tilde{\mathbf{h}}_j\tilde{\mathbf{h}}_j^H - \tilde{\mathbf{h}}_j E\{\Delta\mathbf{h}_j^H \mid \hat{\mathbf{H}}\} - E\{\Delta\mathbf{h}_j \mid \hat{\mathbf{H}}\}\tilde{\mathbf{h}}_j^H] + \sum_{j=1}^{k_{N_T}} E\{\Delta\mathbf{h}_j\Delta\mathbf{h}_j^H \mid \hat{\mathbf{H}}\}\Big\}E_b$$
$$+\sum_{j=k_1}^{k_{i-1}}[\hat{\mathbf{h}}_j\hat{\mathbf{h}}_j^H - \hat{\mathbf{h}}_j E\{\Delta\mathbf{h}_j^H \mid \hat{\mathbf{H}}\} - E\{\Delta\mathbf{h}_j \mid \hat{\mathbf{H}}\}\hat{\mathbf{h}}_j^H]\sigma_{a_j}^2 + \sigma_w^2\mathbf{I}_{N_R}. \quad (176)$$

Based on results discussed in the previous Section, it is evidently that $\Delta\mathbf{h}_j$ is only correlated with $\hat{\mathbf{h}}_j$. Then, we have

$$E\{\Delta\mathbf{h}_j\Delta\mathbf{h}_j^H \mid \hat{\mathbf{H}}\} = E\{\Delta\mathbf{h}_j\Delta\mathbf{h}_j^H \mid \hat{\mathbf{h}}_j\}. \quad (177)$$

From the basic relationship between the autocorrelation and covariance functions, we have

$$E\{\Delta\mathbf{h}_j\Delta\mathbf{h}_j^H \mid \hat{\mathbf{h}}_j\} = Cov\{\Delta\mathbf{h}_j\Delta\mathbf{h}_j^H \mid \hat{\mathbf{h}}_j\} + E\{\Delta\mathbf{h}_j \mid \hat{\mathbf{h}}_j\}E\{\Delta\mathbf{h}_j^H \mid \hat{\mathbf{h}}_j\}. \quad (178)$$

Substituting (171) and (172) into (178), we can write

$$E\{\Delta\mathbf{h}_j\Delta\mathbf{h}_j^H \mid \hat{\mathbf{h}}_j\} = \sigma_{\Delta h}^2\mathbf{I}_{N_R} - \sigma_{\Delta h}^4(\mathbf{R}_{\mathbf{r}} + \sigma_{\Delta h}^2\mathbf{I}_{N_R})^{-1} + \sigma_{\Delta h}^4(\mathbf{R}_{\mathbf{r}} + \sigma_{\Delta h}^2\mathbf{I}_{N_R})^{-1}\mathbf{h}_j\mathbf{h}_j^H(\mathbf{R}_{\mathbf{r}} + \sigma_{\Delta h}^2\mathbf{I}_{N_R}). \quad (179)$$

Introduce the following notations

$$\mathbf{\Lambda} = (\mathbf{R}_{\mathbf{r}} + \sigma_{\Delta h}^2\mathbf{I}_{N_R})^{-1}, \quad (180)$$

$$\mathbf{\Xi} = \mathbf{I}_{N_R} - \sigma_{\Delta h}^2\mathbf{\Lambda}, \quad (181)$$

$$\mathbf{R}_{aa} = diag\{\sigma_{a_{k_1}}^2, \sigma_{a_{k_2}}^2, \ldots, \sigma_{a_{k_{i-1}}}^2\}. \quad (182)$$

Substituting (171) and (179) into (176) and taking into consideration (180)–(182), we have

$$E\{\tilde{\mathbf{s}}_{k_i}\tilde{\mathbf{s}}_{k_i}^H\} = E_b\mathbf{\Xi}\hat{\mathbf{H}}_{k_i:k_{N_T}}\hat{\mathbf{H}}_{k_i:k_{N_T}}^H\mathbf{\Xi} + E_b\sigma_{\Delta h}^4\mathbf{\Lambda}\hat{\mathbf{H}}_{k_1:k_{i-1}}\hat{\mathbf{H}}_{k_1:k_{i-1}}^H\mathbf{\Lambda} + (\hat{\mathbf{H}}_{k_1:k_{i-1}}\mathbf{R}_{aa}\hat{\mathbf{H}}_{k_1:k_{i-1}}^H + N_T E_b\sigma_{\Delta h}^2)\mathbf{\Xi}$$
$$-\sigma_{\Delta h}^2\mathbf{\Lambda}\hat{\mathbf{H}}_{k_1:k_{i-1}}\mathbf{R}_{aa}\hat{\mathbf{H}}_{k_1:k_{i-1}}^H + \sigma_w^2\mathbf{I}_{N_R}, \quad (183)$$

where the notation $\mathbf{H}_{n:m}$ denotes the submatrix containing the n-th to m-th columns of the matrix $\mathbf{H}$.

Based on similar manipulations, we can write

$$E\{2a_{k_i}\tilde{\mathbf{s}}_{k_i}^H \mid \hat{\mathbf{H}}\} - E\{\tilde{\mathbf{s}}_{k_i}\tilde{\mathbf{s}}_{k_i}^H \mid \hat{\mathbf{H}}\} + E\{\mathbf{w}_{AF_i}\mathbf{w}_{AF_i}^H\} = E_b\hat{\mathbf{h}}_{k_i}^H\sigma_{\Delta h}^2(\mathbf{R}_\mathbf{r} + \sigma_{\Delta h}^2\mathbf{I}_{N_R})^{-1} + (\mathbf{w}_{AF}\hat{\mathbf{h}}_{k_i}^H\mathbf{w}_{AF}^H - \mathbf{w}_{PF}\hat{\mathbf{h}}_{k_i}^H\mathbf{w}_{PF}^H)\mathbf{I}_{N_R}. \quad (184)$$

In the root mean-square sense, the second term in (184) representing the GR back end background noise tends to approach zero. By this reason, finally we can write

$$E\{2a_{k_i}\tilde{\mathbf{s}}_{k_i}^H \mid \hat{\mathbf{H}}\} - E\{\tilde{\mathbf{s}}_{k_i}\tilde{\mathbf{s}}_{k_i}^H \mid \hat{\mathbf{H}}\} + E\{\mathbf{w}_{AF_i}\mathbf{w}_{AF_i}^H\} = E_b\hat{\mathbf{h}}_{k_i}^H\sigma_{\Delta h}^2(\mathbf{R}_\mathbf{r} + \sigma_{\Delta h}^2\mathbf{I}_{N_R})^{-1}. \quad (185)$$

Combining (183) and (185), we obtain the MMSE GR output $\widetilde{\mathbf{Y}}_i$, conditionally on $\hat{\mathbf{H}}$.

4.4.3. Computation of LLR

By applying $\widetilde{\mathbf{Y}}_i$ to $\widetilde{\mathbf{s}}_{k_i}$, we have the process at the MMSE GR output [13,52,61] $\widetilde{\mathbf{Z}}_{k_i} = \widetilde{\mathbf{Y}}_i\widetilde{\mathbf{s}}_{k_i}$. According to the Gaussian approximation of the MMSE GR back end process, we can write

$$\tilde{\mathbf{Z}}_{k_i} \approx \tilde{\boldsymbol{\mu}}_{k_i}a_{k_i}^2 + \tilde{\boldsymbol{\eta}}_{k_i}, \quad (186)$$

where

$$\widetilde{\boldsymbol{\mu}}_{k_i} = E\{\widetilde{\mathbf{Y}}_i(\hat{\mathbf{h}}_{k_i} - \Delta\mathbf{h}_{k_i}) \mid \mathbf{H}\} = \widetilde{\mathbf{Y}}_i\boldsymbol{\Xi}\hat{\mathbf{h}}_{k_i} \quad (187)$$

and $\widetilde{\boldsymbol{\eta}}_{k_i} = \mathbf{w}_{AF_{ki}}^2 - \mathbf{w}_{PF_{ki}}^2$ is the background noise at the MMSE GR output with zero-mean and variance $\sigma_{\tilde{\eta}_{k_i}}^2 = 4\sigma_w^4$.

Therefore, the LLR value of the coded bit $\mathbf{g}_{k_i}^\lambda$ can be approximated as [68, 69]

$$\mathscr{L}(\mathbf{g}_{k_i}^\lambda) \approx \frac{1}{\sigma_{\tilde{\eta}_{k_i}}^2}\Big(\min_{\alpha_i \in \mathscr{F}_\lambda^0} |\mathbf{Z}_i - \boldsymbol{\mu}_i\alpha_i|^2 - \min_{\alpha_i \in \mathscr{F}_\lambda^1} |\mathbf{Z}_i - \boldsymbol{\mu}_i\alpha_i|^2\Big), \quad (188)$$

where $\mathscr{F}_\lambda^0$ and $\mathscr{F}_\lambda^1$ denote the modulation constellation symbols subset of $\mathscr{F}$ whose λ-th bit equals 0 and 1, respectively.

4.4.4. Remarks

When the channel estimation error is neglected, i.e., $\sigma_{\Delta h}^2 = 0$ in (180), (181) and (183), the MMSE GR output given by (174) reduces to that of the modified soft-output MMSE GR, in which only decision error propagation is considered, [13, 61]. On the other hand, if $\mathbf{R_r} = \mathbf{I}_{N_R}$ and no residual interference cancellation error is assumed the MMSE GR output given by (174) reduces to that of [50]. For the sake of simplicity, we call this detector as the conventional soft-output MMSE GR hereafter if this detector is applied in channel coded MIMO DS-CDMA wireless communication system. Meanwhile, if both decision error propagation and channel estimation error are neglected, the MMSE GR output given by (174) reduces to that of the conventional MMSE GR output of [51].

4.5. SER definition

We continue a discussion of SER formula derivation presented in (61)-(72), subsection 3.3.1. In the case of *M*-ary PSK system the exact expression for the SER takes the following form [79]

$$P_{SER} = \frac{1}{\pi} \int_0^{\pi - \frac{\pi}{M}} \exp\left\{ -\frac{\frac{E_b}{N_0} \sin^2 \frac{\pi}{M}}{\sin^2 \theta} \right\} d\theta. \tag{189}$$

Taking into account (61), (67), and (189), we can write the *SER* of QPSK system employed the GR in the following form:

$$P_{SER} = \frac{1}{\pi} \int_0^{\pi - \frac{\pi}{M}} \phi_{q_{QBHS/MRC}} \left(-\frac{1}{2 \sin^2 \theta} \right) d\theta. \tag{190}$$

There is a need to note that a direct comparison of QPSK and BPSK systems on the basis of average symbol-energy-to-noise-spectral-density ratio indicates that the QPSK is approximately 3 dB worse than the BPSK.

Another signaling scheme that allows multiple signals to be transmitted using quadrature carries is the QAM. In this case, the transmitted signal can be presented in the following form:

$$a_k(t) = \sqrt{\frac{2}{T_s}} [A_i \cos(2\pi f_c t) + B_i \sin(2\pi f_c t)], \quad 0 \le t \le T_s \tag{191}$$

where A_i and B_i take on the possible values $\pm p; \pm 3p, \ldots, \pm(\sqrt{M}-1)p$ with equal probability, where M is an integer power of 4; T_s is the sampling interval, and f_c is the carrier frequency. The parameter p can be related to the average symbol energy E_b as given by

$$p = \sqrt{\frac{3E_b}{2(M-1)}}. \tag{192}$$

Taking into consideration a definition of the *SER* derived in [80] for *M*-ary QAM system employed the GR, we obtain

$$P_{SER} = 1 - \frac{1}{M}\left\{(\sqrt{M}-2)^2\left[1-2Q\left(\sqrt{\frac{3\varphi_{q_{QBHS/MRC}}}{M-1}}\right)\right]^2 + 4(\sqrt{M}-2)\left[1-2Q\left(\sqrt{\frac{3\varphi_{q_{QBHS/MRC}}}{M-1}}\right)\right]\left[1-Q\left(\sqrt{\frac{3\varphi_{q_{QBHS/MRC}}}{M-1}}\right)\right]\right. \tag{193}$$
$$\left. +4\left[1-Q\left(\sqrt{\frac{3\varphi_{q_{QBHS/MRC}}}{M-1}}\right)\right]^2\right\},$$

where $Q(x)$ is the Gaussian *Q*-function given by

$$Q(x) = \frac{1}{\sqrt{2\pi}}\int_x^\infty \exp(-0.5t^2)dt. \tag{194}$$

4.6. Simulation results

4.6.1. FIR beamforming and MIMO wireless communication system

For a definition of numerical results using simulation, we consider the typical urban channel [81] of the GSM/EDGE system as a practical example. As is usually done for GSM/EDGE, the transmit pulse shape is modeled as a linearized Gaussian minimum-shift keying pulse [82]. The GR input linear system filter is a square-root raised-cosine filter with roll-off factor 0.3. Furthermore, we assume $N_T = 3$ transmit and $N_R = 3$ receive antennas and a maximum channel length of $L = 5$. The correlation coefficient between all pairs of transmit antennas is $\rho = 0.5$.

Figure 9 shows the average $\overline{SNR}$ as a function of the *SNR* noted by E_b/N_0 for the GR with MMSE linear equalization in the cases of FIR (the curves 2 and 3) and infinite impulse response (the curve 1) beamforming filter, respectively, where E_b denotes the average received energy per symbol. The curve 5 corresponds to the case for infinite impulse response beamforming filter for receiver discussed in [67]. The $\overline{SNR}$ was obtained by averaging the respective *SNR*s over 1000 independent realizations of the typical urban channel. For this purpose, in the case of FIR beamforming filter at the GR, the *SNR* given by (159) was used and the corresponding beamforming filters at the GR were calculated using the gradient algorithm discussed in Section 4.3. For infinite impulse response beamforming filter at the GR the result given in [52] was used.

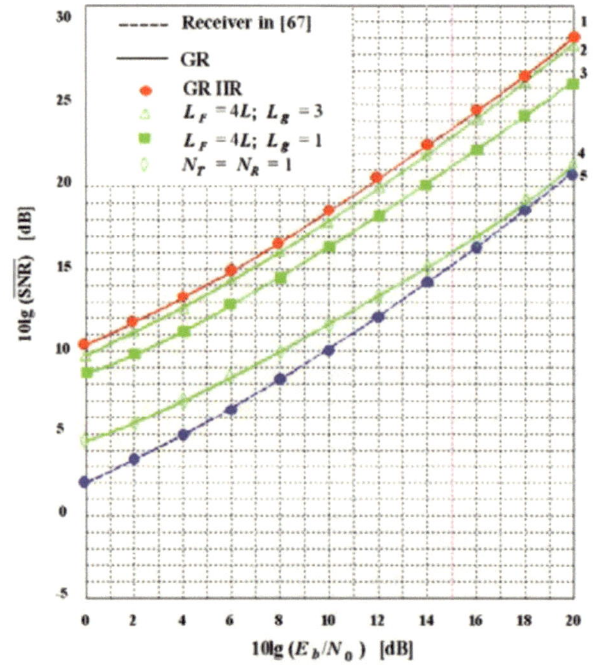

Figure 9. Average *SNR* of MMSE linear equalization at the GR for beamforming with FIR and infinite impulse response filters. The result for single antenna transmission is also indicated (the curve 4). IIR – infinite impulse response beamforming filter.

For comparison, we also show simulation results with FIR linear equalization filters at the GR of length $L_F = 4L$ for FIR beamforming filters at the GR with $L_g = 1$ (the curve 3) optimized for infinite impulse response linear equalization filters at the GR. These simulation results confirm that sufficiently long FIR linear equalization filters at the GR closely approach performance of infinite impulse response linear equalization filters at the GR, which are necessary for (178) to be valid. As expected, the beamforming with infinite impulse response beamforming filters at the GR constitutes a natural performance upper bound for the beamforming with FIR beamforming filters at the GR. However, interestingly, for the typical urban channel the FIR beamforming filter of length $L_g = 3$ (the curve 2) is sufficient to closely approach the performance of the infinite impulse response beamforming at the GR (the curve 1).

We note that for high values of E_b/N_0 even an FIR beamforming filter at the GR of length $L_g = 1$ achieves a performance gain of more than 4.5 dB compared to single antenna transmission, i.e. $N_T = N_R = 1$ (the curve 4). Additional simulations not shown here for other GSM/EDGE channel profiles have shown that, in general, the FIR beamforming filter at the GR of length $L_g \leq 6$ is sufficient to closely approach the performance of the infinite impulse response beamforming at the GR. Thereby, the value of L_g required to approach the performance of the infinite impulse response beamforming at the GR seems to be smaller if the channel is less frequency selective.

4.6.2. Channel estimation and spatially correlation

We choose the 0.5 rate Low Density Parity Check (LDPC) code with a block length of 64800 bits, which is also adopted by DVB-S.2 standard [86]. QPSK modulation with Gray mapping is adopted in $N_T = N_R = 4$ MIMO system. Meanwhile, we set $\sigma_{\Delta h}^2 = 1$ $N_T = N_R$, and $E_P = E_b$. The channel is generated with coherence time of $N_P + N_D = 85$ MIMO vector symbol intervals, and then a LDPC codeword is transmitted via 100 channel coherent time intervals for QPSK modulation. For GR spatially correlated MIMO channel the GR array correlation matrix $\mathbf{R_r}$ with the following elements is adopted [68]

$$\begin{cases} \mathbf{R_r}(n,n) = 1, \mathbf{R_r}(m,n) = \mathbf{R_r^*}(m,n), m,n = 1,2,3,4; \\ \mathbf{R_r}(1,2) = \mathbf{R_r}(2,3) = \mathbf{R_r}(3,4) = 0.4290 + 0.7766j; \\ \mathbf{R_r}(1,3) = \mathbf{R_r}(2,4) = -0.3642 + 0.5490j; \\ \mathbf{R_r}(1,4) = -0.4527 - 0.0015j \end{cases} \tag{195}$$

The performance comparison, in terms of *BER*, of the proposed soft-output MMSE GR, the modified soft-output MMSE GR [83], the conventional soft-output MMSE GR [84], and the conventional MMSE GR [85] is presented in Fig.10 for spatially independent MIMO channel and GR receiver spatially correlated MIMO channel. Also, a comparison with the soft-output MMSE V-BLAST detector discussed in [73] is made. The proposed MMSE GR outperforms all the existing schemes with considerable gain, especially for receiver correlation MIMO channel scenario. The underlying reason of this improvement is that the MMSE GR, by taking channel estimation error, decision error propagation and channel correlation into account, can output more reliable LLR to channel decoder. As channel estimation error is the dominant factor influencing the system performance under the lower *SNR* region, it can observed that the *BER* of the conventional soft-output MMSE GR [84] is slightly better than that of the modified soft-output MMSE GR in the case of spatially independent MIMO channel.

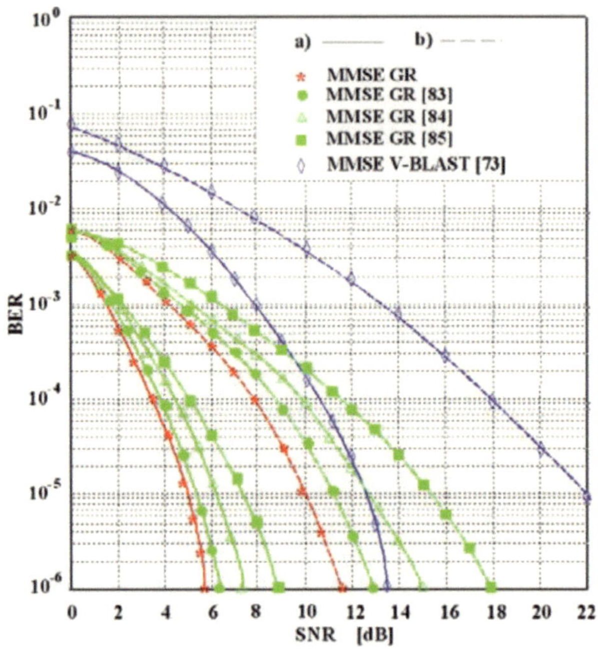

Figure 10. *BER* performance of different detectors under a) spatially independent MIMO channel and b) receiver spatially correlated MIMO channel.

4.7. Conclusions

In the present section, we have considered the FIR beamforming at the GR with perfect channel state information for single carrier transmission over frequency-selective fading channels with zero-forcing linear equalization and GR MMSE linear equalization. We employed a gradient algorithm for efficient recursive calculation of the FIR beamforming filters at the GR. Our results show that for typical GSM/EDGE channel profiles short FIR beamforming filters at the GR suffice to closely approach the performance of optimum infinite impulse response beamforming at the GR discussed in [52]. This is a significant result, since in practice, the quantized beamforming filter coefficients have to be fed back from the receiver to the transmitter, which makes short beamforming filters preferable.

The proposed MMSE GR outperforms all the existing schemes with considerable gain especially for receiver correlation MIMO channel scenario. The underlying reason of this improvement is that the MMSE GR, by taking channel estimation error, decision error propagation, and channel correlation into account, can output more reliable LLR to channel decoder. As channel estimation error is the dominant factor influencing the system performance under the lower *SNR* region, it can observed that the *BER* of the conventional soft-output MMSE GR [87] is slightly better than that of the modified soft-output MMSE GR in the case of spatially independent MIMO channel.

5. Signal processing with fading channels

5.1. Brief review

It is well known that the DS-CDMA transmission technique allows multiple users to share the same spectrum range simultaneously. Using DS-CDMA transmission technique in wireless communication systems, we can reach spectrum efficiency, high system capacity, robustness against interference, high quality of service (QoS) and so on [88,89]. In DS-CDMA wireless communication systems, the concatenating orthogonal Walsh-Hadamard (WH) channelization sequences and pseudonoise (PN) random scrambling sequences are used to generate the orthogonal spreading codes employed by multiple users for simultaneous signal transmission. There are multipaths in DS-CDMA wireless communication systems that destroy orthogonality between codes by introducing non-zero time delays and lead to interference among the transmitted signals in the downlink. One way to solve this problem is to use the scrambling sequences with the purpose to randomize signals transmitted by users and inhomogeneous behavior at nonzero time delays. One technique demonstrating how this problem can be solved is discussed in [90].

At the present time, the wideband DS-CDMA technique employed by wireless communication systems attracts a great attention of researchers, in particular, in mobile communication systems. An important feature of wideband DS-CDMA is an implementation of complex spreading sequences. The term "complex spreading" was generated from terminology used in constant to dual-channel spreading [91]. It is shown in [92-94] that the complex spreading can be realized either by a complex-valued sequence or by two binary sequences. As was shown in [92], the complex sequences can be 3 dB better than the binary Gold sequences [95] in the maximum periodic correlation parameters. Moreover, larger sets are available in complex sequences. Unfortunately, all these sequences are nonorthogonal and can be characterized as the complex-valued PN spreading sequences. Lam and Ozluturk investigated in [95] an application area of nonorthogonal complex sequences in DS-CDMA downlink wireless communication systems. As was defined in [95], the performance bounds were derived for the DS-CDMA wireless communication systems with complex signature sequences over the AWGN channels.

In this section, we investigate a generalized receiver (GR) constructed on the basis of GASP [1-3,6-8]. Using the GR in the DS-CDMA downlink wireless communication system, we would like to get answers on the following questions:

- Is it possible to maintain the orthogonality between users under GR implementation in the DS-CDMA downlink wireless communication system?

- Is it possible to reduce the effect of multipath fading and interference from other users under GR implementation in the DS -CDMA downlink wireless communication system?

- What are benefits under GR implementation in DS-CDMA downlink wireless communication system in comparison with other conventional receivers, for example, the Rake receiver?

To give answers on these questions we carry out our analysis using, for instance, the orthogonal 4-phase complex sequences in the DS-CDMA downlink wireless communication system. These sequences are generated by the unified complex Hadamard transform matrix discussed in [96], the correlation properties of which are studied in [97], where it is shown that the unified complex Hadamard transform sequences possess the better autocorrelation properties in comparison with the WH sequences, which are characterrized by very poor autocorrelation properties.

The use of orthogonal unified complex Hadamard transform sequences by the transmitter as channelization spreading codes scrambled by long PN sequences and further processing these sequences by the GR allows us to maintain the orthogonality between the users, and at the same time, to reduce the effect of multipath fading and interference from other users. A coherent GR [6], for example, can be used to combat the adverse effects of short-term multipath fading in mobile radio propagation environments. Owing to computational simplicity of the signal-to-interference-plus-noise ratio (SINR) in comparison with the probability of error, SINR is mostly used for evaluating and selecting code sequences among several candidates. Therefore, in this section, we investigate the SINR at the GR output when the unified complex Hadamard transform spreading sequences are generated by transmitter in the DS-CDMA downlink wireless communication system and compare this with the SINR at the GR output under transmission of WH real sequences. It is shown that the SINR at the GR output is independent of the phase offsets between different paths when the unified complex Hadamard transform spreading sequences are generated by the transmitter in the DS-CDMA downlink wireless communication system. The SINR at the GR output of the same system is a function of the squared cosine of path phase offsets under generation of WH real sequences by the transmitter. Because of this, as a direct result, the bit error rate (BER) performance of GR employing by the DS-CDMA downlink wireless communication system when the unified complex Hadamard transform spreading sequences are generated by the transmitter is better that that of the system with the WH sequences under Gaussian approximation. Also, we carry out a BER performance comparison of the DS-CDMA system employing the GR with the same system using then conventional receiver, for example, the Rake receiver [95]. Comparative analysis shows us a great superiority in the BER performance under GR employment in the DS-CDMA downlink wireless communication system over the use of the Rake receiver.

This section is organized as follows. At first, we present some basic definitions of the unified complex Hadamard transform sequences. Additionally, we study the DS-CDMA downlink wireless communication system model under the GR employment when the unified complex Hadamard transform spreading sequences are generated and propagated in a multipath fading channel. Further, we investigate the SINR performance at the GR output when the unified complex Hadamard transform spreading sequences are generated and compare this with the SINR of the same system using the WH sequences. Computer simulation and comparison with the Rake receiver are also presented. Some concluding remarks are given.

5.2. Unified complex Hadamard transform sequences

The considered DS-CDMA downlink wireless communication system uses the orthogonal unified complex Hadamard transform spreading sequences. These so-called orthogonal unified complex Hadamard transform sequences are easy to generate. Larger sets of complex sequences are also available. They are categorized into two groups: the half-spectrum property orthogonal unified complex Hadamard transform spreading sequences and the non-half-spectrum property orthogonal unified complex Hadamard transform spreading sequences. Consider briefly how these sequences can be generated and note the main definitions and remarks discussed in [96, 97].

A unified complex Hadamard transform matrix **A** of order $N = 2^n$ is a square matrix with elements ± 1 and $\pm j$, and is constructed by [96,97]

$$A_n = \overset{n}{\underset{i=1}{\otimes}} A_1 = A_1 \otimes A_{n-1} = \underbrace{A_1 \otimes \cdots \otimes A_1}_{n \text{ times}}, \tag{196}$$

where $\otimes$ denotes the Kronecker product, and A_1 is defined as

$$A_1 = \begin{bmatrix} a_1 & a_1 a_3 \\ a_2 & -a_2 a_3 \end{bmatrix} \tag{197}$$

with

$$a_1, a_2, a_3 \in \{1, -1, j, -j\} \quad \text{and} \quad j = \sqrt{-1}. \tag{198}$$

There are 64 different matrices for A_1 satisfying (197) with elements ± 1 and $\pm j$,

$$A_1 A_1^* = A_1^* A_1 = 2I_2 \tag{199}$$

and

$$|\det(A_1)|^2 = 2^2, \tag{200}$$

where $*$ indicates the complex conjugate. Hence, the unified complex Hadamard transform matrix is orthogonal. Furthermore, the unified complex Hadamard transform matrices contain a WH transform matrix as a special case, with

$$a_1 = a_2 = a_3 = 1 \tag{201}$$

in the matrix A_1.

The unified complex Hadamard transform matrices have two categories of 32 basic matrices, depending on whether a_3 in (197) is imaginary or not [97]. If a_3 is imaginary, the matrix group is called the half-spectrum property unified complex Hadamard transform. Otherwise, the group is called the non-half-spectrum property unified complex Hadamard transform. The unified complex Hadamard transform spreading sequence $c_k, k = 1,\ldots,N$ is defined by the k-th row of the unified complex Hadamard transform matrix. It has been shown in [97] that the non-half-spectrum property unified complex Hadamard transform sequences have very similarly poor autocorrelation properties as WH sequences, and some of the half-spectrum property unified complex Hadamard transform sequences exhibit a reasonable compromise between the autocorrelation and cross-correlation functions. In this section, we just consider the half-spectrum property unified complex Hadamard transform sequences, i.e., $a_3 = j \text{ or } -j$.

5.3. System model

In this section, we think that to assess accurately the effects of multipath fading and interference components from other users on the performance of DS-CDMA downlink wireless communication system it is enough to consider a single-cell environment system model. In particular, we analyze a complex baseband-equivalent model with the binary phase-shift keying (BPSK) data and complex signature sequences over a multipath fading channel for the DS-CDMA downlink wireless communication system. The baseband representation of the total signal transmitted on the downlink can be presented in the following form:

$$a(t) = \sum_{k=0}^{K-1} a_k(t) = \sum_{k=0}^{K-1} \sqrt{P_{a_k}}\, b_k(t) c_k(t), \tag{202}$$

where K is the number of users;

$$a_k(t) = \sqrt{P_{a_k}}\, b_k(t) c_k(t) \tag{203}$$

is the transmitted signal of the k-th user; P_{a_k} is the power of the k-th transmitted signal;

$$b_k(t) = \sum_{n=-\infty}^{\infty} b_k^{(n)} p_T(t - nT) \tag{204}$$

is the data signal of the *k*-th user;

$$b_k^{(n)} \in \{-1, +1\} \tag{205}$$

denotes the *n*-th data bit value of the *k*-th user; the function $p_T(t)$ is the rectangular pulse of symbol duration *T*; $c_k(t)$ is the complex spreading signal defined by

$$c_k(t) = \sum_{m=-\infty}^{\infty} c_k^{(m)} \phi(t - mT_c); \tag{206}$$

and $c_k^{(m)}$ denotes the *m*-th complex chip value of the *k*-th user. The function $\varphi(t)$ is a chip waveform that is time-limited to $[0, T_c)$ with

$$\int_0^{T_c} \phi^2(t) dt = T_c, \tag{207}$$

including the rectangular pulse of duration T_c, and T_c is called the chip duration. Throughout this section, we assume that

$$T = NT_c. \tag{208}$$

Power control is assumed to be perfect, and we suppose that the transmitted signal power P_{a_k} is assumed to be known. We also assume

$$c_k^{(m)} = d^{(m)} a_k^{(m)}, \tag{209}$$

where

$$d = \{d^{(m)}\} \text{ with } d^{(m)} \in \{+1, -1\} \tag{210}$$

is the random scrambling code commonly used by all users, and

$$a^{(k)} = \{a_k^{(m)}\} \tag{211}$$

is the user-specific orthogonal unified complex Hadamard transform spreading sequence with period N. Thus,

$$c^{(k)} = \{c_k^{(m)}\} \tag{212}$$

is the random sequence with

$$E\{c_k^{(m)}(c_{k'}^{(n)})^*\} = 0, \quad m \neq n \tag{213}$$

for all k and k', where $E\{\cdot\}$ denotes the mathematical expectation.

The final results of our analysis in this section are applicable to the DS-CDMA downlink wireless communication systems that use long PN scrambling sequences such as m-sequences and Gold sequences [95]. This is because the periods of these long scrambling codes are much larger than that of the spreading factor N, and have correlation properties similar to those of the random scrambling sequences.

For instance, at the base station transmitter in mobile communication system, the signals of all K users are symbol synchronously added before passing through a frequency-selective multipath-fading channel. The complex baseband-equivalent impulse response of the multipath channel can be presented in the following form:

$$h(t) = \sum_{l=0}^{L-1} \alpha_l \exp(j\theta_l)\delta(t - \tau_l), \tag{214}$$

where L is the number of resolvable propagation paths, and $\alpha_l \exp(j\theta_l)$ and τ_l are the complex fading factor and propagation delay of the l-th path, respectively. Note, α_l may be Rayleigh-, Rician-, or Nakagami-distributed, depending on a specific channel model. All random variables in (214) are assumed to be independent for l. The channel parameters, such as α_l, θ_l, and τ_l are here assumed to be known in the dispreading and demodulation process, although in practice, the impulse response of the channel is typically estimated using the pilot symbol or pilot channel.

Moreover, we assume that the multipaths at the GR input are resolvable and chip-synchronized, i.e., they are spaced, at least one chip duration apart in time and the relative delays are multiples of the chip duration. Without loss of generality, the resolved paths are assumed to be numbered such that

$$0 \leq \tau_0 < \tau_1 < \cdots < \tau_{L-1} < T. \tag{215}$$

Hence, the baseband complex representation of the signal at the GR input (the input of GR linear system) of any user is given by

http://dx.doi.org/10.5772/58990

$$x(t) = \sum_{k=0}^{K-1}\sum_{l=0}^{L-1}\sqrt{P_{a_k}}\alpha_l \exp(j\theta_l)b_k(t-\tau_l)c_k(t-\tau_l) + w(t), \tag{216}$$

where $w(t)$ is the complex background AWGN with zero mean and one-sided power spectral density N_0.

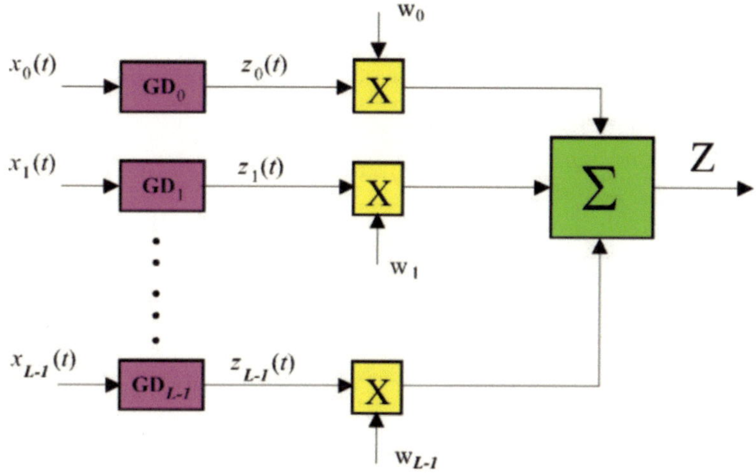

Figure 11. GR structure with $L-1$ fingers.

In order to mitigate the multipath fading effect, the GR with coherent demodulation is implemented. The GR structure is presented in Fig.11, where the number of fingers is equal to the number of resolvable paths.

Since the symbols $b_k^{(n)}$ are i.i.d. from one symbol duration to another and from one user to another, without loss of generality, we focus our attention on the GR output of the user 0 for the zero-th transmitted symbol. The complex GR output of the i-th finger of user 0 in accordance with the GASP in noise [1–3,6–9,11] is

$$z_i = \Re e\left\{\int_{\tau_i}^{T+\tau_i} 2x(t)\sqrt{P_{a_0}^*}c_0^*(t-\tau_i)\exp(-j\theta_i)dt - \int_{\tau_i}^{T+\tau_i} x(t)x^*(t-\tau_i)dt + \int_{\tau_i}^{T+\tau_i}\eta(t)\eta(t-\tau_i)dt\right\}, \tag{217}$$

where $\sqrt{P_{a_0}^*}c_0^*(t-\tau_i)\exp(-j\theta_i)$ is the model of transmitted signal for user 0, i.e., the reference signal generated by the GR, [3,6]; τ_i is the delay factor that can be neglected for simplicity of analysis; $\eta(t)$ is the noise forming at the GR AF output given by (4).

The main functioning condition under employment of GR in DS-CDMA wireless communication systems discussed in detail in Section 2.1 takes the following form

$$P_{a_k} = P^*_{a_k}, \tag{218}$$

where P_{a_k} is the power of information signal and $P^*_{a_k}$ is the power of model signal (the reference signal). In practice, we can perform this matching implementing, for example, tracking systems. These statements and possible ways to solve this problem and how we can implement all this in practice are discussed in detail in [1–3,6–9,11]. Thus, the following process, in a general case, is formed at the GR output for k-th user according to implementation of GASP in DS-CDMA wireless communication systems.

The case 1: a "yes" signal in the input process –

$$\begin{aligned} Z &= \Re e\left\{\int_{\tau_i}^{T+\tau_i}\sum_{l=0}^{L-1} a_k(t)\alpha_l \exp(j\theta_l) a_k^*(t-\tau_i)dt - \int_{\tau_i}^{T+\tau_i}\sum_{l=0}^{L-1}\xi_k(t)\alpha_l \exp(j\theta_l)\xi_k^*(t-\tau_i)dt + \int_{\tau_i}^{T+\tau_i}\sum_{l=0}^{L-1}\eta_k(t)\alpha_l\exp(j\theta_l)\eta_k^*(t-\tau_i)dt\right\} \\ &= \Re e\left\{\int_{\tau_i}^{T+\tau_i}\sum_{l=0}^{L-1}\left[\sqrt{P_{a_k}}\sqrt{P^*_{a_k}}\, b_k(t)\alpha_l\exp(j\theta_l)b_k^*(t-\tau_i)c_k(t)c_k^*(t-\tau_i) - \xi_k(t)\alpha_l\exp(j\theta_l)\xi_k^*(t-\tau_i) + \eta_k(t)\alpha_l\exp(j\theta_l)\eta_k^*(t-\tau_i)\right]dt\right\} \end{aligned}. \tag{219}$$

The case 2: a "no" signal in the input process –

$$Z = \Re e\left\{\int_{\tau_i}^{T+\tau_i}\sum_{l=0}^{L-1}\eta_k(t)\alpha_l\exp(j\theta_l)\eta_k^*(t-\tau_i)dt - \int_{\tau_i}^{T+\tau_i}\sum_{l=0}^{L-1}\xi_k(t)\alpha_l\exp(j\theta_l)\xi_k^*(t-\tau_i)dt\right\} = \Delta_i, \tag{220}$$

where Δ_i is the background noise forming at the GR output. Finally, the GR combiner output that produces a decision statistic is

$$Z = \sum_{i=0}^{L-1} w_i z_i, \tag{221}$$

where the selection of the combining weights w_i's determines the specific diversity-combining technique.

5.4. Performance analysis

5.4.1. SINR at GR output

In this section, we investigate SINR by considering the GR shown in Fig. 11. It follows from (217)-(221) that the i-th GR finger output for user 0 can be presented in the following form

$$z_i = TP_{a_0}\alpha_i b_0^{(0)} + I_{MUI}^{(i)} + I_{MP}^{(i)} + \Delta_i(t), \tag{222}$$

where the first term is the signal component; the second term is the multiple-user interference component determined by

$$I_{MUI}^{(i)} = \Re e\left\{\sum_{k=1}^{K-1} P_{a_k}\alpha_i b_k^{(0)} \int_{\tau_i}^{T+\tau_i} c_k(t-\tau_i)c_0^*(t-\tau_i)dt\right\} \tag{223}$$

the third term is the multipath interference component given by

$$I_{MP}^{(i)} = \Re e\left\{\sum_{k=0}^{K-1}\sum_{l=0,l\neq i}^{L-1} P_{a_k}\alpha_l \exp\left[j(\theta_l-\theta_i)\right] \int_{\tau_i}^{T+\tau_i} b_k(t-\tau_i)c_k(t-\tau_i)c_0^*(t-\tau_i)dt\right\}; \tag{224}$$

and the fourth term $\Delta_i(t)$ is the background noise at the GR output given for a general case by (7).

We can see that the multiple-user interference component $I_{MUI}^{(i)}$ and the multipath interference component $I_{MP}^{(i)}$ are due to the interference from the *i*-th path of other users' signals and from the remaining $L-1$ paths from all users' signals, respectively. The background noise $\Delta_i(t)$ at the GR output is the i.i.d. random process obeying the asymptotic Gaussian distribution with zero mean and variance of $4T\sigma_w^4$ [1–3,6–9,11].

Define a periodic correlation function $R_{k,m}$ by

$$R_c^{(k,m)}(q) = \begin{cases} \sum_{p=0}^{N-1-q} c_k^{(p)}\left(c_m^{(p+q)}\right)^*, & 0 \le q \le N-1 \\ \sum_{p=0}^{N-1+q} c_k^{(p-q)}\left(c_m^{(p)}\right)^*, & 1-N \le q < 0 \\ 0, & |q| \ge N\ . \end{cases} \tag{225}$$

Then, let

$$\tau_l - \tau_i = q_l T_c. \tag{226}$$

With the assumption of chip synchronization, it can be obtained that q_l is an integer and the multiple-user interference and multipath interference components can be presented in the following form:

$$I_{MUI}^{(i)} = \Re e\left\{T_c \sum_{k=1}^{K-1} P_{a_k} \alpha_i b_k^{(0)} R_c^{(k,0)}(0)\right\}, \tag{227}$$

$$I_{MP}^{(i)} = \Re e\left\{T_c \sum_{l=0}^{i-1}\sum_{k=0}^{K-1} P_{a_k} \alpha_l \exp\left[j\left(\theta_l - \theta_i\right)\right] R_{k0} + T_c \sum_{l=i+1}^{L-1}\sum_{k=0}^{K-1} P_{a_k} \alpha_l \exp\left[j\left(\theta_l - \theta_i\right)\right] \hat{R}_{k0}\right\}, \tag{228}$$

where

$$R_{k0} = b_k^{(0)} R_c^{(k,0)}(q_l) + b_k^{(1)} R_c^{(k,0)}(N + q_l), \tag{229}$$

$$\hat{R}_{k0} = b_k^{(-1)} R_c^{(k,0)}(q_l - N) + b_k^{(0)} R_c^{(k,0)}(q_l). \tag{230}$$

In the following, we are going to compare the SINR at the GR output in the DS-CDMA downlink wireless communication system implementing the unified complex Hadamard transform spreading codes with that at the GR output in the DS-CDMA downlink wireless communication system using the WH spreading sequences. Note, that under employment of the orthogonal spreading codes in DS-CDMA downlink wireless communication system, such as the WH real spreading codes and the orthogonal unified complex Hadamard transform spreading codes considered in this section, we must take into consideration that

$$R_c^{(k,0)}(0) = 0\,, \quad k \neq 0. \tag{231}$$

Consequently, the multiple-user interference component is equal to zero, i.e.,

$$I_{MUI}^{(i)} = 0. \tag{232}$$

When the WH spreading codes are used in the DS-CDMA downlink wireless communication system employing the GR, we can obtain that the multipath interference component takes the following form:

$$I_{MP}^{(i)} = T_c \sum_{l=0}^{i-1}\sum_{k=0}^{K-1} P_{a_k} \alpha_l \cos\left(\theta_l - \theta_i\right) R_{k0} + T_c \sum_{l=i+1}^{L-1}\sum_{k=0}^{K-1} P_{a_k} \alpha_l \cos\left(\theta_l - \theta_i\right) \hat{R}_{k0}. \tag{233}$$

Owing to the mutually independent random variables $b_k^{(n)}$ for $0 \le k \le K-1$, the multipath interference component $I_{MP}^{(i)}$ has zero mean. With regard of (213), it can be easily shown via straightforward computation that, for the WH spreading sequences we are able to obtain

$$E\{R_{k0}^2\} = E\{\hat{R}_{k0}^2\} = N. \tag{234}$$

Therefore, the variance of the multipath interference component $I_{MP}^{(i)}$ is denoted by $\sigma^2_{I_{MP}^W}$ and can be determined in the following form:

$$\sigma^2_{I_{MP}^W} = NT_c^2 \sum_{l=0,l\neq i}^{L-1} \alpha_l^2 \cos^2\left(\theta_l - \theta_i\right) \sum_{k=0}^{K-1} P_{a_k}^2. \tag{235}$$

Hence, the SINR at the GR i-th finger output for the WH spreading codes is determined in the following form:

$$SINR_{WH}^{(i)} = \frac{E_{a_0}\alpha_i^2}{\sqrt{\frac{1}{N}\sum_{l=0,l\neq i}^{L-1} \alpha_l^2 \cos^2\left(\theta_l - \theta_i\right) \sum_{k=0}^{K-1} E_{a_k}^2 + 4\sigma_w^4}}, \tag{236}$$

where

$$E_{a_k} = P_{a_k} T, \quad k = 0, 1, \ldots, K-1 \tag{237}$$

is the energy per data symbol of the m-th user and the variance σ_w^2 is given by (23).

Similarly, when the unified complex Hadamard transform spreading codes are employed by the DS-CDMA downlink wireless communication system using the GR, we can obtain that the multipath interference component $I_{MP}^{(i)}$ takes the following form:

$$\begin{aligned} I_{MP}^{(i)} &= T_c \sum_{l=0}^{i-1}\sum_{k=0}^{K-1} P_{a_k}\alpha_l \cos(\theta_l - \theta_i)\mathcal{R}e\{R_{k0}\} + T_c \sum_{l=i+1}^{L-1}\sum_{k=0}^{K-1} P_{a_k}\alpha_l \cos(\theta_l - \theta_i)\mathcal{R}e\{\hat{R}_{k0}\} \\ &+ T_c \sum_{l=0}^{i-1}\sum_{k=0}^{K-1} P_{a_k}\alpha_l \sin(\theta_l - \theta_i)\mathrm{Im}\{R_{k0}\} + T_c \sum_{l=i+1}^{L-1}\sum_{k=0}^{K-1} P_{a_k}\alpha_l \sin(\theta_l - \theta_i)\mathrm{Im}\{R_{k0}\}. \end{aligned} \tag{238}$$

Note that the multipath interference component $I_{MP}^{(i)}$ has zero mean. If the half-spectrum property unified complex Hadamard transform spreading sequences are employed by the DS-CDMA downlink wireless communication system using the GR, it can be easily shown from (213) via straightforward computation that

$$E\left\{\left[\mathcal{R}e(R_{k0})\right]^2\right\} = E\left\{\left[\mathrm{Im}(R_{k0})\right]^2\right\} = \frac{N}{2}, \tag{239}$$

$$E\left\{\left[\mathcal{R}e(\hat{R}_{k0})\right]^2\right\} = E\left\{\left[\mathrm{Im}(\hat{R}_{k0})\right]^2\right\} = \frac{N}{2}. \tag{240}$$

Hence, the variance of the multipath interference component $I_{MP}^{(i)}$ is denoted by $\sigma^2_{I^H_{MP}}$ and can be determined in the following form:

$$\sigma^2_{I^H_{MP}} = \frac{N}{2} T_c^2 \sum_{l=0,l\neq i}^{L-1} \alpha_l^2 \sum_{k=0}^{K-1} P_{a_k}^2. \tag{241}$$

Therefore, the SINR at the GR i-th finger output for the half-spectrum property unified complex Hadamard transform spreading sequences employed by the DS-CDMA downlink wireless communication system using the GR is determined by

$$SINR_H^{(i)} = \frac{E_{a_0}\alpha_i^2}{\sqrt{\frac{1}{2N}\sum_{l=0,l\neq i}^{L-1} \alpha_l^2 \sum_{k=0}^{K-1} E_{a_k}^2 + 4\sigma_w^4}}. \tag{242}$$

5.4.2. GR finger weights

There is a need to note that the interference plus the thermal noise power seen by different fingers of the GR employed by DS-CDMA downlink wireless communication systems is different. Assume that the multipath interference signals are uncorrelated from one finger to another. In this case, the optimal weight in terms of the maximizing SINR at the GR output is dependent on the multipath interference. This optimal weight is called the modified maximal ratio combining (MMRC). There is a need to note that for the traditional maximal ratio combining (MRC) the weights are chosen in the following manner

$$w_i = \alpha_i. \tag{243}$$

Here, if the MMRC is employed as the combiner, the combining weights for DS-CDMA downlink wireless communication system using the WH codes and DS-CDMA downlink wireless communication system with the unified complex Hadamard transform spreading sequences under employment of the GR are different. This is a direct consequence of the difference in the SINR values at the GR finger output of the two systems.

For the WH spreading sequences used by the DS-CDMA downlink wireless communication system employing the GR, MMRC weights take the following form:

$$w_i^{WH} = \sqrt{\frac{P_{a_0}\alpha_i^2}{\frac{1}{N}\sum_{l=0,l\neq i}^{L-1} \alpha_l^2 \cos^2(\theta_l - \theta_i) \sum_{k=0}^{K-1} E_{a_k}^2 + 4\sigma_w^4}}. \tag{244}$$

For the unified complex Hadamard transform spreading sequences used by the DS-CDMA downlink wireless communication system employing the GR, MMRC weights have the following form:

$$w_i^H = \sqrt{\frac{P_{a_0}\alpha_i^2}{\frac{1}{2N}\sum_{l=0,l\neq i}^{L-1}\alpha_l^2\sum_{k=0}^{K-1}E_{a_k}^2 + 4\sigma_w^4}}. \tag{245}$$

After the MMRC scheme, the SINR at the GR output is equal to the sum of all fingers' SINRs, as in (236) or (242). The use of the unified complex Hadamard transform spreading sequences in DS-CDMA downlink wireless communication system employing the GR ensures that the SINR at the GR output is independent of the phase offsets between different paths. The use of the WH real sequences in DS-CDMA downlink wireless communication systems employing the GR causes the SINR at the GR output to be related to the squared cosine of the phase offsets between paths, as seen by comparing (236) and (242). Even though the average SINR per finger at the GR output in the DS-CDMA downlink wireless communication system under employment of the WH codes is the same as that at the GR output in the DS-CDMA downlink wireless communication system under the use of the unified complex Hadamard transform spreading sequences, owing to

$$E\{\cos^2(\theta_l - \theta_i)\} = 0.5, \tag{246}$$

the SINR distribution over the random variable $\cos^2(\theta_l - \theta_i)$ can cause degradation in BER performance under some conditions. This is analogous to the case of a transmission over flat Rayleigh fading channels: even when the Rayleigh gain has a mean of 1, the performance is far worse than in an AWGN channel.

Now, we compare the BER performance of DS-CDMA downlink wireless communication system employing the GR under the use of the unified complex Hadamard transform spreading sequences with that of DS-CDMA downlink wireless communication system employing the GR under the use of the WH real sequences. To analyze the performance of spreading sequences and the diversity-combining schemes considered, we adopt a Gaussian approximation approach based on the central limit theorem [99]. The Gaussian approximation is known not only to give accurate estimations of the probability of error in the region of practical interest, but also to offer insights into the effects of various sequence and system parameters and interference sources on the performance of the GR [1–3]. For simplicity, here we only consider the single finger GR case, i.e., the GR has only one demodulating finger, and the finger is locked onto an arbitrary path, say, the *i*-th multipath component. Under the Gaussian approximation, for the WH spreading sequences a straightforward derivation based on the decision process indicates that the conditional symbol error probability (SEP) for a given

$$\theta = (\theta_0, \theta_1, \ldots, \theta_{L-1}) \tag{247}$$

takes the following form

$$SEP_i^{WH}(\theta) = \Phi\left(\sqrt{SINR_{WH}^{(i)}}\right), \tag{248}$$

where $SINR_{WH}^{(i)}$ is as in (236), and

$$\Phi(x) = \frac{1}{\sqrt{2\pi}} \int_x^{\infty} \exp\left(-\frac{y^2}{2}\right) dy \tag{249}$$

is the error integral.

Averaging (248) with respect to the associated random variables, the average probability of error may be determined in the following form:

$$SEP_i^{WH_{av}} = E_\theta \{SEP_i^{WH}(\theta)\} \tag{250}$$

The averaging may most efficiently be carried out via the Monte Carlo or MatLab techniques.

Under employment of the unified complex Hadamard transform spreading sequences in the DS-CDMA downlink wireless communication system using the GR, the Gaussian approximation leads us to the following form of the conditional SEP:

$$SEP_i^{H} = \Phi\left(\sqrt{SINR_{H}^{(i)}}\right), \tag{251}$$

where $SINR_{H}^{(i)}$ is as in (242). At this point, we should compare (250) and (251). For this purpose, we use a procedure proposed in [99]. If any function $f(x)$ is the convex function and X is the random variable, then Jensen's inequality

$$E\{f(X)\} \geq f(E\{X\}) \tag{252}$$

is satisfied. To apply Jensen's inequality, first define

$$X = \cos^2(\theta_l - \theta_i). \tag{253}$$

Then since θ_i, $i = 0,1,2,\ldots,L-1$ are uniformly distributed within the limits of the interval $[0,2\pi)$, straightforward calculations give us the following result

$$E\{X\} = 0.5. \tag{254}$$

Moreover, the function $f(x)$ here has the following form:

$$f(x) = \Phi\Big(\sqrt{\frac{1}{a+bx}}\Big), \tag{255}$$

where $a > 0$ and $b \geq 0$. Calculating the second derivative of the function $f(x)$ with respect to x, we found that $f(x)$ is a convex function if the following condition is satisfied

$$a + bx \leq \frac{1}{3} \tag{256}$$

is satisfied.

A sufficient condition

$$E_{a_0} \geq \frac{3}{\alpha_i^2}\sqrt{\frac{1}{N}\sum_{l=0,l\neq i}^{L-1}\alpha_l^2\sum_{k=0}^{K-1}E_{a_k}^2 + 4\sigma_w^4} \tag{257}$$

satisfies the inequality (256) and allows applications of Jensen's inequality successively to each component in θ by using (248) and (250). Then we obtain

$$SEP_i^{WH_{av}} \geq \Phi\Big(\sqrt{SINR_{WH}^{(i,low)}}\Big), \tag{258}$$

where

$$SINR_{WH}^{(i,low)} = \frac{E_{a_0}\alpha_i^2}{\sqrt{\frac{1}{2N}\sum_{l=0,l\neq i}^{L-1}\alpha_l^2\sum_{k=0}^{K-1}E_{a_k}^2 + 4\sigma_w^4}}. \tag{259}$$

Comparing the results of (258) and (259) for the WH spreading sequences employed by the DS-CDMA downlink wireless communication system under the use of the GR and the result of (251) for the unified complex Hadamard transform spreading sequences employed by the DS-CDMA downlink wireless communication system under the GR use, we found that

$$SINR_{WH}^{(i,low)} = SINR_{H}^{(i)}. \tag{260}$$

Thus, (251) is a lower bound on the SEP when the WH spreading sequences are used in the DS-CDMA downlink wireless communication system employing the GR if the condition (257) is satisfied. This result implies that the DS-CDMA downlink wireless communication system employing the GR with the unified complex Hadamard transform spreading sequences is more resistant to MAI in comparison with the WH spreading sequences used by the DS-CDMA downlink wireless communication system employing the GR in the case where the only finger is selected in the GR. For more fingers and various combining schemes, although simple closed-form bounds for the SEP of the DS-CDMA downlink wireless communication system employing the GR with the WH spreading sequences are difficult to obtain, we believe that the same conclusion can be made under some similar conditions obtained by using Jensen's inequality and some intensive calculations. This conclusion can be verified under discussion of numerical simulations made in the next section. Furthermore, in view of (58) and (64)–(67), we observe that the DS-CDMA downlink wireless communication system employing the GR with the unified complex Hadamard transform spreading sequences can achieve high reliable performance at not only high SNR E_b/N_0, as in the case of the Rake receiver, but at low SNR, too.

5.5. Simulation results

In this section, we compare the BER performance of the DS-CDMA downlink wireless communication system employing the GR for the cases when the WH spreading sequences and the unified complex Hadamard transform spreading sequences are used under different combining schemes for finger weights such as the traditional equal gain combining (EGC), MRC, and MMRC. Also we present the comparative analysis of BER performance under employment of the GR and Rake receiver by the DS-CDMA downlink wireless communication system. Simulations are performed over the Rayleigh, Ricean, and AWGN channels, respectively.

The spreading sequences of length $N = 64$ are considered, and powers are chosen as

$$P_{a_0} = P_{a_1} = \cdots = P_{a_{K-1}} = 1. \tag{261}$$

Unless stated otherwise, the default system under consideration contains $K = 10$ active users, and the number of paths $L = 4$ in Rayleigh fading multipath channel with

$$E\{\alpha_l^2\} = 1, \; l = 0, 1, \ldots, L-1, \tag{262}$$

and uses the MMRC technique for the GR. In accordance with Section *5.2*, we choose

$$\mu_1 = 1,\ \mu_2 = -j, \quad \text{and} \quad \mu_3 = j \tag{263}$$

to construct a set of the half-spectrum property unified complex Hadamard transform spreading sequences. Note that a different sequence assigned to the user-of-interest results in different BER; the BER in the following examples is an average over the sequence subset.

5.5.1. Effect of GR finger weights

Figure 12 presents different combining techniques in the GR, namely, the traditional EGC, MRC, and MMRC. Additionally, a comparison of the GR with the Rake receiver is made. From presented simulation results it is evident that the GR with MMRC technique has a great superiority over EGC and MRC techniques, especially in the high SNR region. This phenomenon can be explained by the GR finger weights in (244) and (245), where the effect of interference is taken into consideration under MMRC technique, but it is not considered under the traditional EGC and MRC techniques. Additionally, Fig.12 demonstrates a great superiority under employment of the GR by the DS-CDMA downlink wireless communication system with the unified complex Hadamard transform spreading sequences and WH spreading sequences over the implementation of the Rake receiver in the DS-CDMA downlink wireless communication system under the same conditions.

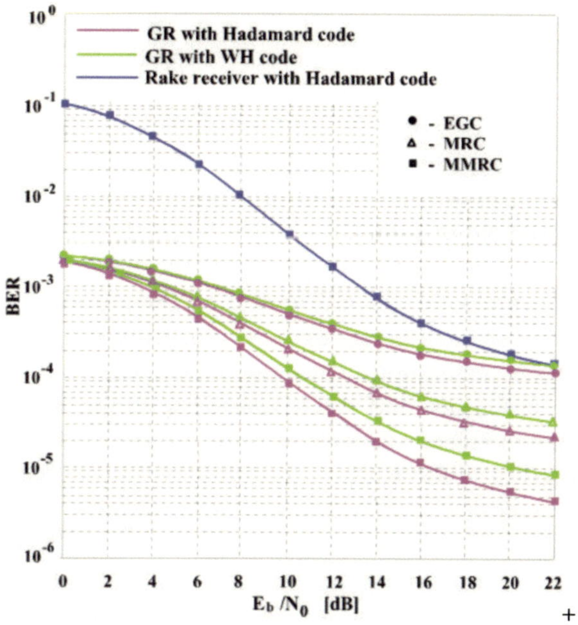

Figure 12. BER performance as a function of SNR E_b/N_0 with different combining techniques in GR over Rayleigh fading channel $(K = 10, L = 4)$.

5.5.2. Effect of different fading channels

Figure 13 shows the BER performance of the DS-CDMA downlink wireless communication system employing the GR with the unified complex Hadamard transform spreading sequences and WH spreading sequences over different channels, including the AWGN channel, Rayleigh fading channel, and Ricean fading channel with the factor $F = 5$. It can be seen that in all the channels under consideration, the DS-CDMA downlink wireless communication system employing the GR with the unified complex Hadamard transform spreading sequences has the better BER performance, especially in the high SNR region. Also, we can see a great superiority under employment of the GR in the DS-CDMA downlink wireless communication system with the unified complex Hadamard transform spreading sequences and WH spreading sequences over the implementation of the Rake receiver in the DS-CDMA downlink wireless communication system under the same conditions.

5.5.3. Effect of MAI

To show the effect of MAI, Fig.14 depicts the BER performance of the DS-CDMA downlink wireless communication system employing the GR with the unified complex Hadamard transform spreading sequences and WH spreading sequences as a function of the number of active users *K*. The BER performance of the DS-CDMA downlink wireless communication system employing the GR with the unified complex Hadamard transform spreading sequences is better than that of the DS-CDMA downlink wireless communication system employing the GR with the WH spreading sequences at the high SNR and all the chosen values of *K*. Although the advantage brought by using the unified complex Hadamard transform spreading sequences in the DS-CDMA downlink wireless communication system employing the GR fades gradually with the increase of MAI. For example, under the large number of users and small SNR, the BER performance of the DS-CDMA downlink wireless communication system employing the GR with the unified complex Hadamard transform spreading sequences is comparable to that of the DS-CDMA downlink wireless communication system employing the GR with the WH spreading sequences in the low SNR region. Also, for all cases, a great superiority of the GR implementation in the DS-CDMA downlink wireless communication system with the unified complex Hadamard transform spreading sequences and WH spreading sequences in comparison with employment of the Rake receiver in the same system is evident. These results in simulations have verified our theoretical analysis.

5.6. Conclusions

In this section, we investigate a possibility to employ both the orthogonal unified complex Hadamard transform spreading sequences and the WH real spreading sequences in the DS-CDMA downlink wireless communication system using the GR constructed based on the GASP and carry out a comparative analysis. The SINR at the GR output in the DS-CDMA

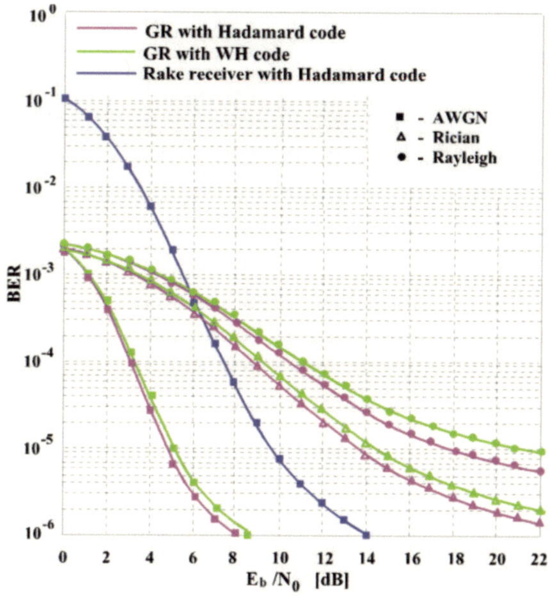

Figure 13. BER performance as a function of SNR E_b/N_0 in different channels $(K=10, L=4)$, MMRC.

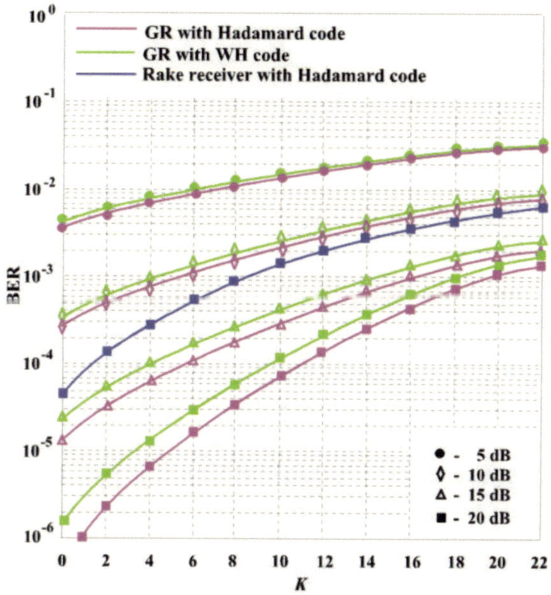

Figure 14. BER performance as a function of the number of users K. Rayleigh fading channel, MMRC, $L=4$.

downlink wireless communication system with the unified complex Hadamard transform spreading sequences is independent of the phase offsets between different paths. The SINR at the GR output in the DS-CDMA downlink wireless communication system with the WH real spreading sequences is a function of the squared cosine of path phase offsets. By this reason, the employment of the orthogonal unified complex Hadamard transform spreading sequences in the DS-CDMA downlink wireless communication system using the GR provides the better BER performance with respect to the DS-CDMA downlink wireless communication system using the GR with the WH real spreading sequences, especially, at high SNRs. Evaluation of the BER performance demonstrating the benefits of employment of the unified complex Hadamard transform spreading sequences and MMRC weights in the GR employed by the DS-CDMA downlink wireless communication system has been carried out over simulation. Comparative analysis of employment of the GR and Rake receiver in the DS-CDMA downlink wireless communication system with the orthogonal unified complex Hadamard transform spreading sequences and the WH real spreading sequences has been performed over simulation. Comparison showed a great superiority in favor of the GR.

Author details

Vyacheslav Tuzlukov

Department of Communications and Information Technologies, School of Electronics Engineering, College of IT Engineering, Kyungpook National University, Daegu, South Korea

References

[1] Tuzlukov V. A new approach to signal detection theory. *Digital Signal Processing*, 1998; 8 (3) pp. 166 –184.

[2] Tuzlukov V. *Signal Processing in Noise: A New Methodology*, Minsk: IEC, 1998.

[3] Tuzlukov V. *Signal Detection Theory*, New York: Springer-Verlag, 2001.

[4] Shbat M.S., Tuzlukov V. Spectrum sensing under correlated antenna array using generalized detector in cognitive radio systems. *International Journal of Antennas and Propagation*, 2013, Vol. 2013, Article ID 853746, 8 pages, 2013. doi:10.1155/2013/853746.

[5] Shbat M.S., Tuzlukov V. Evaluation of detection performance under employment of the generalized detector in radar sensor systems. *Radioengineering*. 2014; 23 (1) pp. 50–65.

[6] Tuzlukov V. *Signal Processing Noise*, Boca Raton, London, New York, Washington D.C.: CRC Press, Taylor & Francis Group, 2002.

[7] Tuzlukov V. *Signal and Image Processing in Navigational Systems*. Boca Raton, London, New York, Washington D.C.: CRC Press, Taylor & Francis Group, 2005

[8] Tuzlukov V. *Signal Processing in Radar Systems*, Boca Raton, London, New York, Washington D.C.: CRC Press, Taylor & Francis Group, 2012

[9] Tuzlukov V. Generalized Approach to Signal Processing in Wireless Communications: The Main Aspects and Some Examples. Chapter 11 in *Wireless Communications and Networks: Recent Advances*, Editor: Eksim Ali, INTECH, Chroatia, 2012, pp.305– 338.

[10] Tuzlukov V. – Editor. *Communication Systems: New Research*. NOVA Science Publishers, Inc., New York, USA, 2013, 423 pp.

[11] Tuzlukov V. Wireless Communications: Generalized Approach to Signal Processing., Chapter 6 in *Communication Systems: New Research*. Editor: Vyacheslav Tuzlukov. NOVA Science Publishers, Inc., New York, USA, 2013, pp. 175–268.

[12] Tuzlukov V. Optimal combining, partial cancellation, and channel estimation and correlation in DS-CDMA systems employing the generalized detector. *WSEAS Transactions on Communications*, 2009; 8 (7) pp. 718–733.

[13] Tuzlukov V. Multiuser generalized detector for uniformly quantized synchronous CDMA signals in AWGN channels. *Telecom munications Review*. 2010; 20 (5) pp. 836–848.

[14] Tuzlukov V. Signal processing by generalized detector in DS-CDMA wireless communication systems with frequency-selective channels," *Circuits, Systems and Signal Processing*. 2011; 30 (6) pp. 1197–1230.

[15] Tuzlukov V. Signal processing by generalized receiver in DS-CDMA wireless communication systems with optimal combining and partial cancellation. *EURASIP Journal on Advances in Signal Processing*. 2011, Vol. 2011, Article ID 913189, 15 pages, doi:10.1155/2011/913189; 2011.

[16] Tuzlukov V. DS-CDMA downlink systems with fading channel employing the generalized receiver. *Digital Signal Processing*, 2011; 21 (6) pp. 725–733.

[17] Tuzlukov V. Design of optimal waveforms in MIMO radar systems based on the generalized approach to signal processing," *WSEAS Transactions on Communications*. 2012; 11 (12) pp. 448–462.

[18] Tuzlukov V. Implementation of generalized detector in MIMO radar systems. *WSEAS Transactions on Communications*, 2013;12 (3) pp. 107–120.

[19] Tuzlukov V. Bit error probability of quadriphase DS– CDMA wireless communication systems based on generalized approach to signal processing. *Telecommunications Review*. 2013; 23 (4) pp. 501–515.

[20] Tuzlukov V. Error probability performance of quadriphase DS-CDMA wireless communication systems based on generalized approach to signal processing. *WSEAS Transactions on Communications*. 2014; 13 (2) pp. 116–129.

[21] Shbat M.S., Tuzlukov V. Definition of adaptive detection threshold under employment of the generalized detector in radar sensor systems," *IET Signal Processing*. 2014; 8 (6) pp.622–632.

[22] Maximov M. Joint correlation of fluctuative noise at the outputs of frequency filters. *Radio Engineering and Telecommunications*.1956; 35 (9) pp. 28–38.

[23] Chernyak Y. Joint correlation of noise voltage at the outputs of amplifiers with non-overlapping responses. *Radio Physics and Electronics*. 1960; 28 (4) pp. 551–561.

[24] Shbat M.S. Performance analysis of signal detection by GD in radar sensor and cognitive radio systems," *Kyungpook National University, PhD Thesis*, December 2013.

[25] Taboga M. *Lectures on Probability Theory and Mathematical Statistics*. 2nd Ed. Amazon Create Space. 2012.

[26] Gradshteyn I., Ryzhik I. *Table of Integrals, Series, and Products*, 7th Ed. Academic Press: Amsterdam, Boston, London, New York, Oxford, Paris, San Diego, San Francisco, Singapore, Sydney, Tokyo, 2007.

[27] Shbat M.S, Tuzlukov V. Noise power estimation under generalized detector employment in automotive detection and tracking systems," in *Proceedings of the 9th IET Data Fusion and Target Tracking Conf. (DF&TT'12)*, doi: 10.1049/cp.2012.0416, Lon- don, UK, 2012.

[28] Alouini M.S., Abdi A., Kaveh M. Sum of gamma variates and performance of wireless communications systems over Nakagami-fading channels. *IEEE Transactions on Vehicular Technology*. 2001; 50 (6) pp.1471 - 1480.

[29] Win M.Z., Winters J.H. Analysis of Hybrid Selection/Maximal Ratio Combining in Rayleigh Fading. *IEEE Transactions on Communications*. 1999; 47 (12) pp.1773–1776.

[30] Win M.Z., Winters J.H. Virtual Branch Analysis of Symbol Error Probability for Hybrid Selection/Maximal-Ratio Combining in Rayleigh Fading. *IEEE Transactions on Communications*. 2001; 49 (11) pp.1926–1934.

[31] Kong N., Milstein L.B. Average SNR of a Generalized Diversity Selection Combining Scheme. *IEEE Communications Letters*. 1999; 3 (3) pp.57–59.

[32] Alouini M.S., Simon M.K. Performance of Coherent Receivers with Hybrid SC/MRC over Nakagami Fading Channels," *IEEE Transactions on Vehicular Technology*. 1999; 48 (4) pp.1155–1164.

[33] Alouini M.S., Simon M.K. A MGF-Based on Performance Analysis of Generalized Selection Combining over Rayleigh Fading Channels," *IEEE Transactions on Communications.* 2000; 48 (3) pp.401–415.

[34] Malik R.K., Win M.Z. Analysis of Hybrid Selection/Maximal-Ratio Combining in Correlated Nakagami Fading," *IEEE Transactions on Communications.* 2002; 50 (8) pp.1372–1383.

[35] Annamalai A., Tellambura C. A New Approach to Performance Evaluation of Generalized Selection Diversity Receivers in Wireless Channels," in Proceedings *IEEE Vehicular Technology Conference,* Vol. 4, October 2001, pp. 2309–2313.

[36] Proakis, J.G. *Digital Communications,* 4th ed. New York: McGraw-Hill, 2001.

[37] Mengali U., Andrea A.N.D. *Synchronization Techniques for Digital Receivers,* New York: Plenum, 1997.

[38] Tong L., Perreau S. Multichannel Blind Identification: From Subspace to Maximal Likelihood Methods. *Proceedings IEEE.* 1998; 86 (10) pp.1951–1968.

[39] Tugnait J.K, Tong L., Ding Z. Single-User Channel Estimation and Equalization. *IEEE Signal Processing Magazine.* 2000; 17 (3) pp.17–28.

[40] Abed-Meraim K., Qaiu W., Hua Y. Blind System Identification. *Proceedings IEEE.* 1997; 85 (8) pp.1310–1322.

[41] Alouini M.S., Goldsmith A.J. A Unified Approach for Calculating Error Rates of Linearly Modulated Signals over Generalized Fading Channels. *IEEE Transactions on Communications.* 1999; 47 (9) pp.1324–1334.

[42] Tuzlukov V., Yoon W.S., Kim Y.D. Wireless Sensor Networks Based on the Generalized Approach to Signal Processing with Fading Channels and Receive Antenna Array. *WSEAS Transactions on Circuits and Systems.* 2004; 10 (3) pp. 2149–2155.

[43] Divsalar D., Simon M.K., Raphaeli D. Improved Parallel Interference Cancellation for CDMA.. *IEEE Transactions on Communications.* 1998; 46 (2) pp.258–268.

[44] Renucci P.G., Woerner B.C. Optimization of Soft Interference Cancellation for DS-CDMA. *Electronics Letters.* 1998; 34 (4), pp.731–733.

[45] Guo D Linear Parallel Interference Cancellation in CDMA, Master Degree Dissertation in Engineering. *National University of Singapore,* 1998.

[46] Hsieh Y.-T., Wu W.-R. Optimal Two-Stage Decoupled Partial PIC Receivers for Multiuser Detection. *IEEE Transactions on Wireless Communications.* 2005; 4 (1) pp.112-127.

[47] Ghotobi M., Soleymani M.R. Multiuser Detection of DS-CDMA Signals Using Partial Parallel Interference Cancellation in Satellite Communications. *IEEE Journal on Selective Areas in Communications.* 2004; 22 (4) pp.584-593.

[48] Wu K.-W., Wang C.-L. Soft-Input Soft-Output Partial Parallel Interference Cancellation for DS-CDMA Systems. in *Proceed- ings IEEE International Conference on Communications (ICC'01),* Helsinki, Finland, June 2001, pp.1172–1176.

[49] Papoulis A., Pillai S.U. *Probability, Random Variables and Stochastic Processes,* 4th ed, New York: McGraw-Hill, 2002.

[50] Xue G., Weng, J.F., Lee-Ngoc T., Tahar S An Analytic Model for Performance Evaluation of Parallel Interference Cancellers in CDMA Systems. *IEEE Communications Letters.* 2000; 4 (6) pp.84–186.

[51] Kim J.H., Tuzlukov V., Yoon W.-Y., Kim Y.D. Generalized Detector under Non-orthogonal Multipulse Modulation in Remote Sensing Systems. *WSEAS Transactions on Signal Processing.* 2005; 2 (1) pp.203–208.

[52] Tuzlukov V., Yoon W.S., Kim Y.D. Adaptive Beam-Former Generalized Detector in Wireless Sensor Networks. in *Proceedings of the IASTED International Conference on Parallel and Distributed Computing and Networks (PDCN 2004),* February 17-19, Innsbruck, Austria, 2004, pp.195–200.

[53] Kim J.H., Tuzlukov V., Yoon W.-Y., Kim Y.D. Macrodiversity in Wireless Sensor Networks Based on the Generalized Approach to Signal Processing. *WSEAS Transactions on Communications.* 2005; 8 (4) pp.648–653.

[54] Craig J.W. A New, Simple and Exact Result for Calculating the Probability of Error for Two-Dimensional Signal Constellations. in *Proceedings IEEE MILCOM'91 Conference,* November15-18, Boston, MA, USA, 1991, pp. 25.5.1–25.5.5.

[55] Ziemer R.E., Tranter W.H. *Principles of Communications: Systems, Modulation, and Noise.* 6th Ed. New York, NY: John Wiley & Sons, Inc., 2010.

[56] Renucci P.G., Woerner B.D. Optimization of Soft Interference Cancellation for DS-CDMA. *Electron. Letters.* 1998; 34 (4) pp. 731–733.

[57] Morrow Jr. R.K., Lehnert J.S. Bit-to-Bit Error Dependence in Slotted DS/SSMA Packer Systems with Random Signature Sequences. *IEEE Transactions on Communications.* 1989; 37 (10) pp.1052–1061.

[58] Liberti Jr. J.C., Rappaport T.S. *Smart Antennas for Wireless Communications: IS-95 and Third Generation CDMA Applications.* Upper Saddle River, NJ: Prentice Hall, 1999.

[59] Brown III D.R., Johnson Jr. C.R SINR, Power Efficiency, and Theoretical System Capacity of Parallel Interference Cancellation *Journal of Communications and Networks.* 2001; 3 (9) pp.228–237.

[60] Manohar S., Tikiya V., Annavajjala R., Chockalingam A. BER-Optimal Linear Parallel Interference Cancellation for Multicarrier DS-CDMA in Rayleigh Fading. *IEEE Transactions on Communications*. 2007; 55 (6) pp.1253–1265.

[61] Tuzlukov V. Selection of Partial Cancellation Factors in DS-CDMA Systems Employing the Generalized Detector," in *Proceed ings 12th WSEAS International Conference on Communications: New Aspects of Communications*, July 23-25, Heraklion Crete, Greece, 2008, pp.129–134.

[62] Dighe P., Mallik R., Jamuar S. Analysis of Transmit-Receive Diversity in Rayleigh Fading. *IEEE Transactions on Communications*. 2003; 51(4) pp.694–703.

[63] Jongren G., Skoglund M., Ottersten B. Combining Beamforming and Orthogonal Space-Time Block Coding, *IEEE Transactions on Information Theory*. 2002; 48(3) pp.611–627.

[64] Love D., Heath R., Strohmer T. Grassmannian Beamforming for Multiple-Input Multiple-Output Wireless Systems, *IEEE Transactions on Information Theory*. 2003; 49(10) pp.2735–2747.

[65] Mukkavilli K., Sabharwal A., Erkip E., Aazhang B. Beamforming with Finite Rate Feedback in Multiple Antenna Systems. *IEEE Transactions on Information Theory*. 2003; 49(10) pp.2562–2579.

[66] Choi J., Heath R. Interpolation Based Transmit Beamforming for MIMO-OFDM with Limited Feedback. *IEEE Transactions on Signal Processing*. 2005; 53(11) pp.4125–4135.

[67] Palomar D.P., Lagunas M.A. Joint Transmit-Receive Space-Time Equalization in Spatially Correlated MIMO Channels: a Beamforming Approach. *IEEE Journal on Selected Areas in Communications*. 2003; 21 pp.730–743.

[68] Lee H., Lee B., Lee I. Iterative Detection and Decoding with an Improved V-BLAST for MIMO-OFDM Systems, *IEEE Journal on Selected Areas in Communications*. 2006; 24(3) pp.504 –513.

[69] Lee H., Lee I. New Approach for Error Compensation in Coded V-BLAST OFDM Systems. *IEEE Transactions on Communications*. 2007; 55(2) pp.345–355.

[70] Wang J., Wen O.Y., Li S. Capacity and Performance of MIMO BICM System with Soft-Output MMSE Soft Interference Cancellation. in *Proceedings CCNC 2008*, 2008, January 10-12, Las Vegas, Nevada, USA, pp.100–104.

[71] Marzetta T.L. BLAST Training: Estimating Channel Characteristics for High-Capacity Space-Time Wireless, in *Proceedings 37th Annual Allerton Conference on Communications, Control, and Computing*. 1999 pp.958–966.

[72] Taricco G., Biglieri E. Space-Time Decoding with Imperfect Channel Estimation. *IEEE Transactions on Wireless Communications*. 2005; 4(4) pp.1874–1888.

[73] Lee K., Chun J. Symbol Detection in V-BLAST Architectures under Channel Estimation Error. *IEEE Transactions on Wireless Communications*. 2007; 6(2) pp.593–597.

[74] Clark M.V., Greenstein L.J., Kennedy W.K., Shafi M. Optimum Linear Diversity Receivers for Mobile Communications, *IEEE Transactions on Vehicular Technology*. 1994; 43(2) pp. 47–56.

[75] Cioffi J.M., Dudevoir G.P., Eyuboglu M.V., Forney Jr. G.D. MMSE Decision-Feedback Equalizers and Coding - Part I: Equalization results. *IEEE Transactions on Communications*. 1995; 43(10) pp.2582–2594.

[76] Ghotobi M., Soleymani M.R. Multiuser Detection of DS-CDMA Signals Using Partial Parallel Interference Cancellation in Satellite Communications. *IEEE Journal on Selected Areas in Communications*. 2004; 22(4) pp. 584–593.

[77] T. K. Moon and W. C. Stirling, *Mathematical Methods and Algorithms for Signal Processing*. New York, Prentice Hall, 2000.

[78] Kay S.M., *Fundamentals of Statistical Signal Processing: Estimation Theory*. Englewood Cliffs, NJ: Prentice -Hall, 1993.

[79] Craig J.W. A New, Simple and Exact Result for Calculating the Probability of Error for Two-Dimensional Signal Constellations, in *Proceedings IEEE International Conference on Military Communications, Milcom'91*, November 4-7, 1991, McLean, Virgi- nia, USA, Vol. 2, pp.571–575, 1991.

[80] Ziemer R.E., Tranter W.H. *Principles of Communications: Systems, Modulation, and Noise*. 6th Ed. John Wiley & Sons, Inc., New York, 2010.

[81] GSM recommendation 05.05: Propagation conditions, Version 5.3.0, release 1996.

[82] Mukkavilli K., Sabharwal A., Erkip E., Aazhang B. Beamforming with Finite Rate Feedback in Multiple Antenna Systems, *IEEE Transactions on Information Theory*. 2003; 49(10) pp.2562–2579.

[83] Tuzlukov V. Multiuser Generalized Detector for Uniformly Quantized Synchronous CDMA Signals in Wireless Sensor Networks with Additive white Gaussian Noise Channels, in *Proceedings International Conference on Control, Automation, and Systems (ICCAS 2008)*, October 14-17, 2008, Seoul, Korea, pp.1526–1531.

[84] Tuzlukov V. Selection of Partial Cancellation Factors in DS-CDMA Systems Employing the Generalized Detector, in *Proceedings 12th WSEAS International Conference on Communications: New Aspects of Communications*, July 23-25, 2008, Heraklion, Crete, GREECE, 2008, pp.129–134.

[85] Kim J.H, Tuzlukov V., Yoon W.S, Kim Y.D. Performance Analysis under Multiple Antennas in Wireless Sensor Networks Based on the Generalized Approach to Signal Processing. *WSEAS Transactions on Communications*, 2005; 4 (7) pp.391–395.

[86] ETSI EN302 307 V1.1.1 (2004-06), Digital Video Broadcasting (DVB); Second Generation Framing Structure, Channel Coding and Modulation Systems for Broadcasting, Interactive Services, News Gathering and Other Broad-Band Satellite Applications.

[87] Tuzlukov V., Yoon W.S., Kim Y.D. MMSE Multiuser Generalized Detector for Non-orthogonal Multipulse Modulation in Wireless Sensor Networks, in *Proceedings 9th World Multi-Conference on Systemics, Cybernetics and Informatics (WMSCI 2005),* July 10-13, 2005, Orlando, Florida, USA, 2005.

[88] Lee J.S., Miller L.E. *CDMA Systems Engineering Handbook.* Boston, MA: Artech House, 1998.

[89] Pickholtz R.L., Milstein L.B., Schiling D.L. Spread Spectrum for Mobile Communications, *IEEE Transactions on Vehicular Technology.* 1991; 40 (2) pp.313–322.

[90] Fong M. H., Bhargava V.K., Wang Q. Concatenated Orthogonal/PN Spreading Sequences and Their Application to Cellular DS- CDMA systems with Integrated Traffic. *IEEE Journal on Selected Areas in Communications.* 1996; 14(3) pp.547–558.

[91] Peterson R.L., Ziemer R.E, Borth D.E. *Introduction to Spread-Spectrum Communication.* Englewood Cliffs, NJ: Prentice-Hall, 1995.

[92] Boztas S., Hammons R., Kumar P.V. 4-Phase Sequences with Near-Optimum Correlation Properties, *IEEE Transactions on In- formation Theory.* 1992; 38(3) pp.1101–1113.

[93] Chu D.C. Polyphase Codes with Good Periodic Correlation Properties. *IEEE Transactions on Information Theory.* 1972;18 (3) pp.531–532.

[94] Frank R.L., Zadoff S.A. Phase Shift Pulse Codes with Good Periodic Correlation Properties. *IEEE Transactions on Information Theory.* 1962; 8 (2) pp.381–382.

[95] Lam A.W., Ozluturk F.M. Performance Bounds for DS/SSMA Communications with Complex Signature Sequences. *IEEE Transactions on Communications.* 1992; 40 (10) pp.1607–1614.

[96] Rahardja S., Falkowski B.J. Family of Unified Complex Hadamard Transforms. *IEEE Transactions on Circuits and Systems– II: Analog and Digital Signal processing.* 1999; 46 (8) pp.1094–1100.

[97] Rahardja S., Ser W., Lin Z. UCHT-Based Complex Sequences for Asynchronous CDMA Systems. *IEEE Transactions on Communications.* 2003; 51 (5) pp.618–626.

[98] Proakis J.G. *Digital Communications.* 4th Ed. New York: McGraw-Hill, 2001.

[99] Borovkov A.A. *Probability Theory.* Amsterdam, The Netherlands: Gordon and Breach, 1998.

Chapter 5

Free Space Optical Communications — Theory and Practices

Abdulsalam Ghalib Alkholidi and
Khaleel Saeed Altowij

Additional information is available at the end of the chapter

http://dx.doi.org/10.5772/58884

1. Introduction

1.1. FSO concepts

1.1.1. What is Free Space Optics (FSO)?

FSO is a line-of-sight technology that uses lasers to provide optical bandwidth connections or FSO is an optical communication technique that propagate the light in free space means air, outer space, vacuum, or something similar to wirelessly transmit data for telecommunication and computer networking. Currently, FSO is capable of up to 2.5 Gbps [1] of data, voice and video communications through the air, allowing optical connectivity without requiring fiber-optic cable or securing spectrum licenses. Operate between the 780 – 1600 nm wavelengths bands and use O/E and E/O converters. FSO requires light, which can be focused by using either light emitting diodes (LEDs) or lasers (light amplification by stimulated emission of radiation). The use of lasers is a simple concept similar to optical transmissions using fiber-optic cables; the only difference is the transmission media. Light travels through air faster than it does through glass, so it is fair to classify FSO as optical communications at the speed of the light. FSO communication is considered as an alternative to radio relay link line-of sight (LOS) communication systems. This chapter is concentrate on ground-to-ground free-space laser communications. FSO components are contain three stages: transmitter to send of optical radiation through the atmosphere obeys the Beer-Lamberts`s law, free space transmission channel where exist the turbulent eddies (cloud, rain, smoke, gases, temperature variations, fog and aerosol) and receiver to process the received signal. Typical links are between 300 m and 5 km, although longer distances can be deployed such as 8–11 km are possible depending on the speed and required availability. The importance of this chapter is to introduce the FSO technique step by step.

We will briefly focus on concept of FSO technology in section 1. Section 2 presents an optical wireless transceiver design and FSO main components and transmission media. Mathematical model of atmospheric turbulence of FSO is illustrated in section 3. Second part of this study is a case study to adapt between theoretical and practical parts of FSO technique, where series of simulations results are demonstrated and analyzed. In section 4, we demonstrate the first practical part, simulation results and discussion of geometric loss and total attenuation. The second part of case study explores the optical link budget is presented in section 5. Third part of case study shows the simulation results of BER and SNR of this proposed work is demonstrated in section 6. Section 7 presents some concluding remarks. Finally, we propose some important questions related to this chapter for self-evaluation.

1.1.2. FSO applications [1,2]

- Telecommunication and computer networking
- Point-to-point LOS links
- Temporary network installation for events or other purpose as disaster recovery
- For communications between spacecraft, including elements of satellite constellation
- Security applications
- Military application: (its potential for low electromagnetic emanation when transferring sensitive data for air forces)
- Metro network extensions: carriers can deploy FSO to extend existing metropolitan area fiber rings, to connect new networks, and, in their core infrastructure, to complete SONET rings.
- Enterprise connectivity: the ease with which FSO links can be installed makes them a natural for interconnecting local area network segments that are housed in buildings separated by public streets or other right-of-way property.
- Fiber backup: FSO may also be deployed in redundant links to backup fiber in place of a second fiber link.
- Backhaul: FSO can be used to carry cellular telephone traffic from antenna towers back to facilities wired into the public switched telephone network.
- Service acceleration: FSO can be also used to provide instant service to fiber-optic customers while their fiber infrastructure is being laid.
- Last-Mile access: In today's cities, more than 95% of the buildings do not have access to the fiber optic infrastructure due to the development of communication systems after the metropolitan areas. FSO technology seems a promising solution to the connection of end-users to the service providers or to other existing networks. Moreover, FSO provides high-speed connection up to Gbps, which is far more beyond the alternative systems.

http://dx.doi.org/10.5772/58884

The advantages and disadvantages of FSO are as following [1,2]:

1.1.3. FSO Advantages

- Long distance up to 8 km.
- High bit rates speed rates: the high bandwidth capability of the fiber optic of 2.5 Gbps to 10 Gbps achieved with wavelength division multiplexing (WDM). Modern systems can handle up to 160 signals and can thus expand a basic 10 Gbit/s system over a signal fiber pair to over 1.6 Tbit/s.
- Immunity from electromagnetic interference: secure cannot be detected with RF meter or spectrum analyzer, very narrow and directional beams
- Invisible and eye safe, no health hazards so even a butterfly can fly unscathed through a beam
- Low bit error rates (BER)
- Absence of side lobes
- Deployment of FSO systems quickly and easily
- No Fresnel zone necessary
- Low maintenance (Practical)
- Lower costs as compared to fiber networks (FSO costs are as low as 1/5 of fiber network costs).
- License-free long-range operation (in contrast with radio communication)

1.1.4. FSO disadvantages

For terrestrial applications, the principal limiting factors are:

Beam dispersion, atmospheric absorption, rain, fog, snow, interference from background light sources (including the sun), shadowing, pointing stability in wind, and pollution.

1.1.5. Comparison between FSO vs. fiber optics vs. other technologies

In the future fiber optics replaced by FSO for the following reasons:

- Optics is the study of the behavior and properties of light
- Optical fibers can carry a laser beam for long distances
- Most of the recent large effort of digging up the ground and laying down new fiber has been directed towards extending the fiber optic backbone to new central offices, and not laying fiber directly to the customer
- Like fiber, FSO uses lasers to transmit data, but instead of enclosing the data stream in a glass fiber, it is transmitted through the air.

1.2. Light and electromagnetic spectrum

The electromagnetic spectrum is the range of all possible frequencies of electromagnetic radiation. The "electromagnetic spectrum" of an object has a different meaning, and is instead the characteristic distribution of electromagnetic radiation emitted or absorbed by that particular object. The electromagnetic spectrum extends from below the low frequencies used for modern radio communication to gamma radiation at the short-wavelength (high-frequency) end, thereby covering wavelengths from thousands of kilometers down to a fraction of the size of an atom. The limit for long wavelengths is the size of the universe itself, while it is thought that the short wavelength limit is in the vicinity of the Planck length, although in principle the spectrum is infinite and continuous. Most parts of the electromagnetic spectrum are used in science for spectroscopic and other probing interactions, as ways to study and characterize matter. In addition, radiation from various parts of the spectrum has found many other uses for communications and manufacturing (see electromagnetic radiation for more applications).

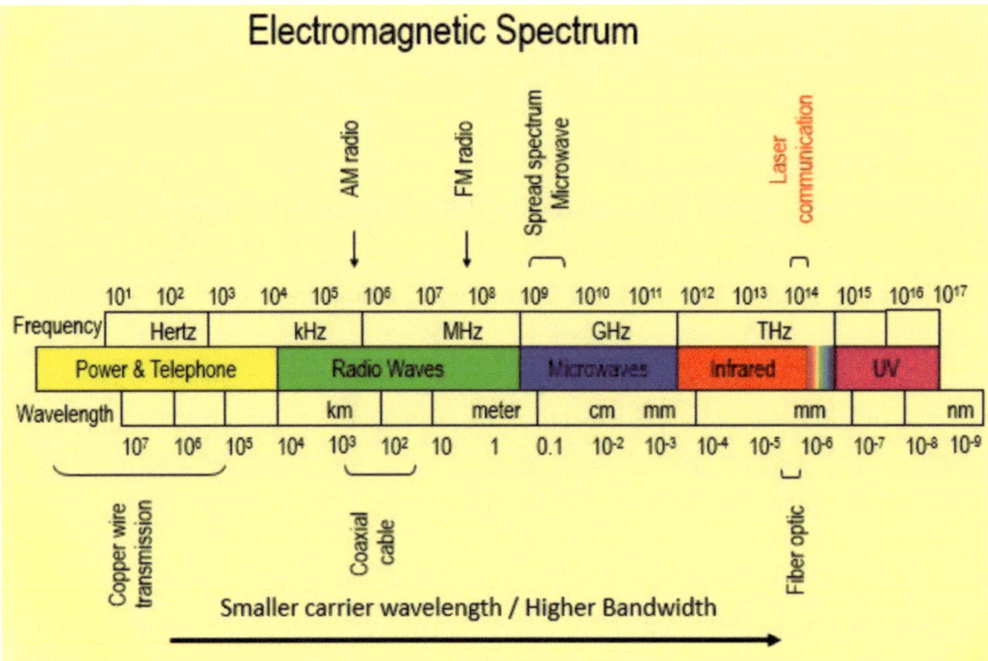

Figure 1. The electromagnetic spectrum.

The electromagnetic spectrum as demonstrated in Fig. 1, can be expressed in term of wavelength, frequency, or energy. Wavelength (λ), frequency (ν) are related by the expression [3]. The higher the frequency, the higher the energy.

$$\lambda = \frac{c}{\nu} \tag{1}$$

http://dx.doi.org/10.5772/58884

Where c is the speed of light ($2.998 \times 10^8\ m / s$). The energy of the various components of the electromagnetic spectrum is given by the expression

$$E = h\nu \tag{2}$$

Where h is Planck`s constant = 6.63×10^{-34} Joule seconds. The units of wavelength are meters with the terms microns (denoted μm and equal to $10^{-6}\ m$) and nanometers ($10^{-9}\ m$) being used just as frequently. Frequency is measured in Hertz (Hz), with one Hertz being equal to one cycle of one cycle of sinusoidal wave per second. A commonly used unit of energy is the electron-volt.

There are several transmission windows that are nearly transparent (attenuation < 0.2 dB/km), between 780 nm and 1600 nm wavelength range. These windows are located around several specific center wavelengths:

- 850 nm

Characterized by low attenuation, the 850 nm window is very suitable for FSO operation. In addition, reliable, high-performance, and inexpensive transmitter and detector components are generally available and commonly used in today's service provider networks and transmission equipment. Highly sensitive silicon avalanche photo diode (APD) detector technology and advanced vertical cavity surface emitting laser (VCSEL) technology can be used for operation in this atmospheric window [4].

- 1060 nm

The 1060 nm transmission window shows extremely low attenuation values. However, transmission components to build FSO system in this wavelength range are very limited and are typically bulky (e.g. YdYAG solid state lasers). Because this window is not specially used in telecommunications systems, high-grade transmission components are rare. Semiconductor lasers especially tuned to the nearby 980 nm wavelength (980 nm pump lasers for fiber amplifiers) are commercially available. However, the 980 nm wavelength range experiences atmospheric attenuation of several dB/km even under clear weather conditions.

- 1250 nm

The 1250 nm transmission window offers low attenuation, but transmitters operating in this wavelength range are rare. Lower power telecommunications grade lasers operating typically between 1280-1310 nm are commercially available. However, atmospheric attenuation increases drastically at 1290 nm, making this wavelength only marginally suitable for free space transmission.

- 1550 nm

The 1550 nm band is well suited for free space transmission due to its low attenuation, as well as the proliferation of high-quality transmitter and detector components. Components include very high-speed semiconductor laser technology suitable for WDM operation as well as amplifiers (EDFA, SOA) used to boost transmission power. Because of the attenuation properties and component availability at this range, development of WDM free space optical systems is feasible.

1.3. Laser principles

A laser is similar in function to an LED, but somewhat different both in how it functions and in its characteristics. The idea of stimulated emission of radiation originated with Albert Einstein around 1916. Until that time, physicists had believed that a photon could interact with an atom only in two ways: The photon could be absorbed and raise the atom to a higher energy level, or the photon could be emitted by the atom when it dropped into a lower energy level [5].

Figure 2 shows the typical energy diagram (term scheme) of an atom. An electron can be moved into a higher energy level by energy provided from the outside. As a basic rule, not all transitions are allowed, and the time that an electron stays in a higher energy state before it drops to a lower energy level varies. When the electron drops from a higher to a lower level, energy is released. A radiative transition that involves the emission of a photon in the visible or infrared spectrum requires a certain amount of energy difference between both energy levels.

Figure 2. Energy level diagram.

$$I = \frac{c}{\left|E_i - E_j\right| / h} = \frac{1.2398}{E_i - E_h} \tag{3}$$

For ease of understanding, we will describe laser operation by using only two energy levels. Figure 3 illustrates the different methods of photon interaction [5].

http://dx.doi.org/10.5772/58884

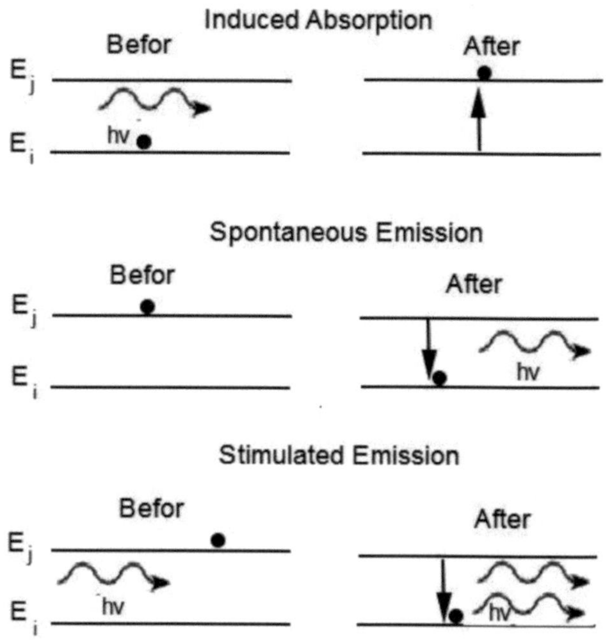

Figure 3. Understanding laser operation.

There are three possibilities:

- Induced absorption: an incoming photon whose wavelength matches the difference between the energy levels E_j and E_i can be absorbed by an atom that is in the lower energy state. After this interaction process, the photon disappears, but its energy is used to raise the atom to an upper energy level.

- Spontaneous emission: an atom in the upper energy level can spontaneously drop to the lower level. The energy that is released during this transition takes the form of an emitted photon. The wavelength of the photon corresponds to the energy difference between the energy states E_j and E_i. This resembles the process of electron-hole recombination, which resulted in the emission of a photon in the LED structure. Gas-filled fluorescent lights operate through spontaneous emission.

- Stimulated emission: an atom in the upper level can drop to the lower level, emitting a photon with a wavelength corresponding to the energy difference of the transition process. The actual emission process is induced by an incoming photon whose wavelength matches the energy transition level of the atom. The stimulated photon will be emitted in phase with the stimulating photon, which continues to propagate.

When these three processes take place in a media such as a solid-state material or gas-filled tube, many atoms are involved. If more atoms are in the ground state (or lower excited level) than in the upper one, the number of photons entering the material will decrease due to absorption.

However, if the number of photons in the upper level exceeds the number of photons in the lower level, a condition called population inversion is created. Laser operation requires the state of population inversion because under these circumstances, the number of photons increases as they propagate through the media due to the fact that more photons will encounter upper-level atoms than will meet lower-level atoms. Keep in mind that upper-level atoms cause the generation of additional photons, whereas lower-level atoms would absorb photons. A medium with population inversion has gain and has the characteristics of an amplifier.

A laser is a high-frequency generator, or oscillator. To force the system to oscillate, it needs amplification, feedback, and a tuning mechanism that establishes the oscillation frequency. In a radio-frequency system, such feedback can be provided by filtering the output signal with a frequency filter, connecting the output signal back to the input, and electronically amplifying the signal before it is coupled back into the input stage. In the case of a laser, the medium provides the amplification. Therefore, a medium capable of laser operation is often referred to as active media. For more details about fundamental of FSO technology, readers merely can refer to reference [5], chapter 2.

1.3.1. Laser diodes

The entire commercial free-space optics industry is focused on using semiconductor lasers because of their relatively small size, high power, and cost efficiency. Most of these lasers are also used in fiber optics; therefore, availability is not a problem. From the semiconductor design point of view, two different laser structures are available: edge emitting lasers and surface-emitting lasers. With an edge emitter, the light leaves the structure through a small window of the active layer and parallel to the layer structure. Surface emitters radiate through a small window perpendicular to the layer structure.

Edge emitters can produce high power. More than 100 milliwatts at modulation speeds higher than 1 GHz are commercially available in the 850 nm wavelength range. The beam profile of edge-emitting diodes is not symmetrical. A typical value for this elliptical radiation output pattern is 20 × 35 degrees. This specific feature can cause a problem when the output power has to be coupled efficiently into a fiber and external optics such as cylindrical lenses are used to increase the coupling efficiency. Surface-emitting diodes typically produce less power output. However, the beam pattern is close to being symmetrical or round. A typical value for the beam divergence angle is 12 degrees. This feature is beneficial for coupling light into a (round) optical fiber. Besides discussing basic designs of semiconductor lasers, we will also provide information regarding WDM laser sources and look into Erbium Doped Fiber Amplifiers/lasers that have been discussed recently for use in FSO systems.

1.3.2. Basic designs of optical lenses

A lens is a piece of glass or other transparent material that refracts light rays in such a way that they can form an image. Lenses can be envisioned as a series of tiny refracting prisms, and

http://dx.doi.org/10.5772/58884

each of these prisms refracts light to produce its own image. When the prisms act together, they produce an image that can be focused at a single point.

Lenses can be distinguished from one another in terms of their shape and the materials from which they are made. The shape determines whether the lens is converging or diverging. The material has a refractive index that determines the refractive properties of the lens. The horizontal axis of a lens is known as the principal axis. A converging (convex) lens directs incoming light inward toward the center axis of the beam path. Converging lenses are thicker across their middle and thinner at their upper and lower edges. When collimated[1] (parallel) light rays enter a converging lens, the light is focused to a point. The point where the light converges is called the focal point and the distance between the lens and the focal point is called focal length. A diverging (convex) lens directs incoming rays of light outward away from the axis of the beam path. Diverging lenses are thinner across their middle and thicker at their upper and lower edges. Figure 4 illustrates the behavior of converging and diverging lenses [6].

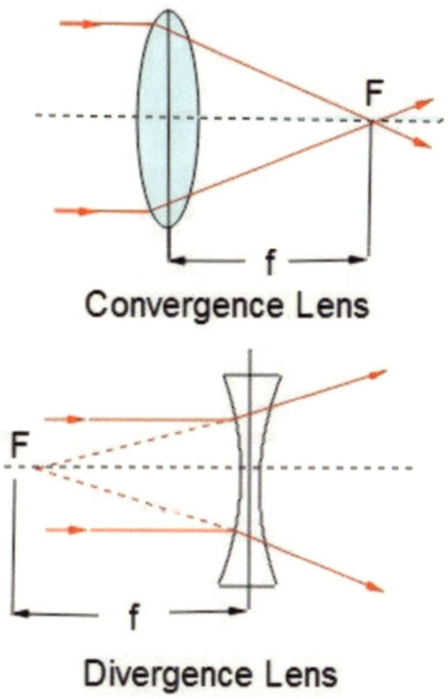

Figure 4. Converging and diverging lenses.

The focal length (*f*) of an optical system is a measure of how strongly the system converges or diverges light. For an optical system in air, it is the distance over which initially collimated rays are brought to a focus. A system with a shorter focal length has greater optical power than one with a long focal length; that is, it bends the rays more strongly, bringing them to a focus in a shorter distance. The focal length *f* is then given by

1 Make (rays of light or particles) accurately parallel: (as adjective collimated) a collimated electron beam.

$$\frac{1}{f} = \frac{1}{u} + \frac{1}{v} \qquad (4)$$

where u is the distance between the light source and the lens, and v is the distance between the lens and the screen.

1.4. Important definitions

After illustrating the basic concepts of FSO, we return to the important definitions related to the laser power reduction due to atmospheric channel effects phenomena. These definitions are considered as the core principle of FSO transmission channel turbulence namely atmosphere, aerosol, absorption, scattering, and radiance etc. Absorption and scattering are related to the loss and redirection of the transmitted energy. The majority of these definitions will be discussed in detail in the case study of this chapter (section 4).

An atmosphere is a layer of gases surrounding a planet or other material body material of sufficient mass that is held in place by the gravity of the body. An atmosphere is more likely to be retained if the gravity is high and the atmosphere's temperature is low. Earth atmospheric, which is mostly nitrogen, also contains oxygen used by most organism for respiration and carbon dioxide used by plants, algae and cyanobacteria for photosynthesis, also protects living organisms from genetic damage by solar ultraviolet radiation. Another definition of an atmosphere is the envelope of gases surrounding the earth or another planet.

An aerosol is defined as a colloidal system of solid or liquid particles in a gas. An aerosol includes both the particles and the suspending gas, which is usually air. This term describes an aero-solution, clouds of microscopic particles in air. According to the literature, the size range of aerosol particles to be only from 0.1 to 1 μm another authors indicate that the size of aerosol is between 0.01 and 10 μm in radius. Another definition of aerosol is extremely-fine liquid droplets or solid particles that remain suspended in air as **fog** or **smoke**.

Fog is a thick cloud of tiny water droplets suspended in the atmosphere at or near the earth's surface that obscures or restricts visibility (to a greater extent than mist; strictly, reducing visibility to below 1 km).

Smoke is a visible suspension of carbon or other particles in air, typically one emitted from a burning substance.

Haze is traditionally an atmospheric phenomenon where dust, smoke and other dry particles obscure the clarity of the sky.

Dust is a fine powder made up of very small pieces of earth or sand.

Absorption of the light is the decrease in intensity of optical radiation (light) as it passes through a material medium owing to its interaction with the medium. In the process of absorption, the energy of the light is converted to different forms of internal energy of the medium; it may be completely or partially re-emitted by the medium at frequencies other than the frequency of the absorbed radiation.

Light scattering is a form of scattering in which light is the form of propagating energy which is scattered. Light scattering can be thought of as the deflection of a ray from a straight path, for example by irregularities in the propagation medium, particles, or in the interface between two media. Deviations from the law of reflection due to irregularities on a surface are also usually considered to be a form of scattering. When these are considered to be random and dense enough that their individual effects average out, this kind of scattered reflection is commonly referred to as diffuse reflection. Scattering has different types as Rayleigh, Mie, Tyndall, Brillion, and Raman Scattering.

Radiance erasures of the quantity of radiation that passes through or is emitted from a surface and falls within a given solid angle in a specified direction. Radiance is also used to quantify emission of neutering and other particles.

Radiance (in Watts): total amount of energy that flows the light source.

Attenuation is the gradual loss in intensity of any kind of flux through a medium. Attenuation affects the propagation of waves and signals transmission media.

Scintillation is a flash of light produced in a transparent material by the passage of a particle (an electron, an alpha particle an ion, or a high-energy photon).

The process of scintillation is one of luminescence whereby light of a characteristic spectrum is emitted following the absorption of radiation. The emitted radiation is usually less energetic than that absorbed. Scintillation is an inherent molecular property in conjugated and aromatic organic molecules and arises from their electronic structures. Scintillation also occurs in many inorganic materials, including salts, gases, and liquids.

1.5. Lasers and eye safety

According to reference [5], certain high-power laser beams used for medical procedures can damage human skin, but the part of the human body most susceptible to lasers is the eye. Like sunlight, laser light travels in parallel rays. The human eye focuses such light to a point on the retina, the layer of cells that responds to light. Like staring directly into the sun, exposure to a laser beam of sufficient power can cause permanent eye injury.

For that reason, potential eye hazards have attracted considerable attention from standards writers and regulators. The standards rely on parameters such as laser wavelength, average power over long intervals, peak power in a pulse, beam intensity, and proximity to the laser. Laser wavelength is important because only certain wavelengths—between about 400 nm and 1,550 nm—can penetrate the eye with enough intensity to damage the retina. The amount of power the eye can safely tolerate varies with wavelength. This is dominated by the absorption of light by water (the primary component in the eye) at different wavelengths.

The vitreous fluid of the eye is transparent to wavelengths of 400—1,400 nm. Thus, the focusing capability of the eye causes approximately a 100,000-to-1 concentration of the power to be focused on a small spot of the retina. However, in the far infrared (1,400 nm and higher), such light is not transmitted by the vitreous fluid, so the power is less likely to be transferred to the retina. Although damage to the corneal surface is a possibility, the focusing capabilities of the eye do not lead to large magnification of power densities. Therefore, much greater power is required to cause damage. The relevance of this is that lasers deployed in FSO that utilize wavelengths greater than 1,400 nm are allowed to be approximately 100 times as powerful as FSO equipment operating at 850 nm and still be considered eye safe. This would be the "killer app" of FSO except that the photo diode receiver technologies suffer reduced sensitivity at greater than 1,400 nm, giving back a substantial portion of the gain. Also, lasers that operate at such wavelengths are more costly and less available. Nevertheless, at least one FSO manufacturer has overcome these obstacles and currently offers equipment deploying multiple 1,550 nm lasers.

With respect to infrared radiation, the absorption coefficient at the front part of the eye is much higher for longer wavelength (> 1,400 nm) than for shorter wavelength. As such, damage from the ultraviolet radiation of sunlight is more likely than from long wavelength infrared. Eye response also differs within the range that penetrates the eyeball (400 nm—1,400 nm) because the eye has a natural aversion response that makes it turn away from a bright visible light, a response that is not triggered by an (invisible) infrared wavelength longer than 0.7 μm. Infrared light can also damage the surface of the eye, although the damage threshold is higher than that for ultraviolet light.

High-power laser pulses pose dangers different from those of lower-power continuous beams. A single high-power pulse lasting less than a microsecond can cause permanent damage if it enters the eye. A low-power beam presents danger only for longer-term exposure. Distance reduces laser power density, thus decreasing the potential for eye hazards.

2. Optical wireless transceiver design

FSO contains three components: transmitter, free space transmitted channel line of sight, and receiver. Transmitter is considered as an optical source 1-laser diode (LD) or 2-light emitting diode (2-LED) to transmit of optical radiation through the atmosphere follows the Beer-Lamberts's law as indicated in subsection 3.6 Eq. 34.

FSO link is demonstrated as in Fig. 5. The selection of a laser source for FSO applications depends on various factors. It is important that the transmission wavelength is correlated with one of the atmospheric windows. As noted earlier, good atmospheric windows are around 850 nm and 1550 nm in the shorter IR wavelength range. In the longer IR spectral range, some wavelength windows are present between 3–5 micrometers (especially 3.5–3.6 micrometers) and 8–14 micrometers [5]. However, the availability of suitable light sources in these longer wavelength ranges is pretty limited at the present moment. In addition, most sources need low temperature cooling, which limits their use in commercial telecommunication applications. Other factors that impact the use of a specific light source include the following:

- Price and availability of commercial components
- Transmission power
- Lifetime

- Modulation capabilities
- Eye safety
- Physical dimensions
- Compatibility with other transmission media such as fiber.

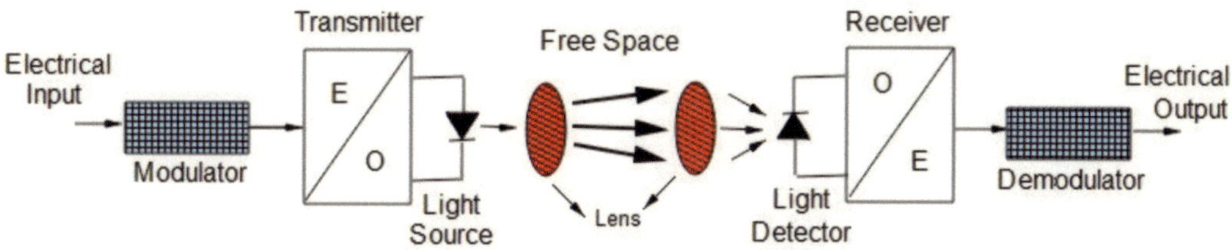

Figure 5. Block diagram of an optical wireless link showing the front end of an optical transmitter and receiver [7].

Electrical input is a network traffic into pulses of invisible light representing 1`s and 0`s. The transmitter, which consists of two part main parts: an interface circuit and source driver circuit, converts the input signal to an optical signal suitable for transmission. The drive circuit of the transmitter transforms the electrical signal to an optical signal by varying the current follow through the light source. Transmitter function is to project the carefully aimed light pulses into the air. This optical light source can be of two types:

1. A light-emitting diode (LED) or
2. A laser diode (LD).

The information signal modulates the field generated by the optical source. The modulated optical field then propagates through a free-space path before arriving at the receiver. In the receiver side, transmitted data realizes inverse operations i.e., photo detector converts the optical signal back into an electrical form as indicated in previous figure. In other words, a receiver at the other end of the link collects the light using lenses and/or mirrors. Received signal converted back into fiber or cooper and connected to the network. Reverse direction data transported the same way (full duplex). We can see, anything that can be done in fiber can be done with FSO.

Equation (5) illustrates the data rate of FSO system:

$$Data\ Rate\left[{}^{bits}\!/_{sec}\right] = \frac{1}{\eta} P_r\left[\frac{photons}{sec}\right] \tag{5}$$

Where P_r is a received power, and η is a received power sensitivity of the receiver [photons/bit].

Small angles – divergence angle and spot size between transmitter and receiver are presented in Fig. 6.

$$1^{\circ} \approx 17\ mrad \rightarrow 1\ mrad \approx 0.0573^{\circ}$$

θ is a divergence angle between transmitter and receiver FSO units.

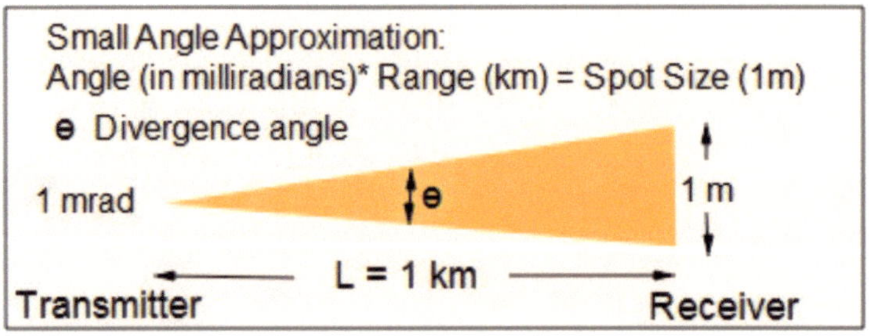

Figure 6. Small angles – divergence and spot size between transmitter and receiver.

The geometric path loss for an FSO link depends on the beam-width of the optical transmitter, the path length (L), and the divergence angle (θ). Transmitter and receiver aperture diameters are quantifiable parameters, and are usually specified by manufacturer. Table (1) illustrates the relation of divergence in (mrad), range in (km), and spot diameter in (inches or feet).

Divergence	Range	Spot Diameter
0.5 mrad	1.0 km	~ 0.5 m (~ 20 in)
2.0 mrad	1.0 km	~ 2.0 m (~ 6.5 ft)
4.0 mrad (~ ¼ deg)	1.0 km	~ 4.0 m (~ 13.0 ft)

Table 1. The divergence, range, and spot diameter.

3. Mathematical model of atmospheric turbulence

The atmospheric attenuation is one of the challenges of the FSO channel, which may lead to signal loss and link failure. The atmosphere not only attenuates the light wave but also distorts and bends it. Transmitted power of the emitted signal is highly affected by scattering and turbulence phenomena. Attenuation is primarily the result of absorption and scattering by molecules and particles (aerosols) suspended in the atmosphere. Distortion, on the other hand, is caused by atmospheric turbulence due to index of refraction fluctuations. Attenuation affects the mean value of the received signal in an optical link whereas distortion results in variation of the signal around the mean.

3.1. Aerosol

Aerosols are particles suspended in the atmosphere with different concentrations. They have diverse nature, shape, and size. Aerosols can vary in distribution, constituents, and concentration. As a result, the interaction between aerosols and light can have a large dynamic, in terms of wavelength range of interest and magnitude of the atmospheric scattering itself. Because most of the aerosols are created at the earth's surface (e.g., desert dust particles, human-made industrial particulates, maritime droplets, etc.), the larger concentration of aerosols is in the boundary layer (a layer up to 2 km above the earth's surface). Above the boundary layer, aerosol concentration rapidly decreases. At higher elevations, due to atmospheric activities and the mixing action of winds, aerosol concentration becomes spatially uniform and more independent of the geographical location. Scattering is the main interaction between aerosols and a propagating beam. Because the sizes of the aerosol particles are comparable to the wavelength of interest in optical communications, Mie scattering theory is used to describe aerosol scattering [8].

Type	Radius (μm)	Concentration (in cm^{-3})
Air molecules	10^{-4}	10^{19}
Aerosol	10^{-2} to 1	10 to 10^{3}
Fog	1 to 10	10 to 100
Cloud	1 to 10	100 to 300
Raindrops	10^{2} to 10^{4}	10^{-5} to 10^{-2}
Snow	10^{3} to $5{\times}10^{3}$	N/A
Hail	$5{\times}10^{3}$ to $5{\times}10^{4}$	N/A

Table 2. Radius ranges for various types of particles.

Such a theory specifies that the scattering coefficient of aerosols is a function of the aerosols, their size distribution, cross section, density, and wavelength of operation. The different types of atmospheric constituents' sizes and concentrations of the different types of atmospheric constituents are listed in Table (2) [7,9].

3.2. Visibility Runway Visual Range (RVR)

Visibility was defined originally for meteorological needs, as a quantity estimated by a human observer. It defined as (Kruse model) means of the length where an optical signal of 550 nm is reduced to 0.02 of its original value [10]. However, this estimation is influenced by many subjective and physical factors. The essential meteorological quantity, namely the transparency of the atmosphere, can be measured objectively and it is called the Runway Visual Range (RVR) or the meteorological optical range [11]. Some values of atmospheric attenuation due to scattering based on visibility are presented in Table (3).

Visibility S (Line of Sight) (km)	λ = 800 nm (dB/km)	λ = 2500 nm (dB/km)
0.5	32.5	30.8
0.7	23	21
0.9	18	16
1.1	14.5	12.5
1.3	12	10
1.5	10	8.33
Source:		

Table 3. Variation in atmospheric attenuation due scattering based on visibility (data obtained from [7,12]).

When the length difference between the two optical paths varies, the energy passes through minima and maxima. The visibility V is defined by:

$$V = \frac{I_{Max} - I_{min}}{I_{Max} + I_{min}} \tag{6}$$

The visibility depends on the degree of coherence of the source, on the length difference between the paths as well as on the location of the detector with respect to the source. The coherence between the various beams arriving at the detector also depends on the crossed media: for example the diffusing medium can reduce the coherence. For links referred to as "in direct sight" links, coherent sources can be used, provided that parasitic reflections do not interfere with the principal beam, inducing modulations of the detected signal [11].

Low visibility will decrease the effectiveness and availability of FSO systems, and it can occur during a specific time period within a year or at specific times of the day. Low visibility means the concentration and size of the particles are higher compared to average visibility. Thus, scattering and attenuation may be caused more in low visibility conditions [13].

3.3. Atmospheric attenuation

Atmospheric attenuation is defined as the process whereby some or all of the electromagnetic wave energy is lost when traversing the atmosphere. Thus, atmosphere causes signal degradation and attenuation in a FSO system link in several ways, including absorption, scattering, and scintillation. All these effects are varying with time and depend on the current local conditions and weather. In general, the atmospheric attenuation is given by the following Beer's law equation [14]:

$$\tau = \exp(-\beta L), \tag{7}$$

where,

τ is the atmospheric attenuation;

β is the total attenuation coefficient and given as

$$\beta = \beta_{abs} \beta_{scat};$$ (8)

L is the distance between transmitter and receiver (unit: km);

β_{abs} is the molecular and aerosol absorption, this parameter value is considered as too small so, we can neglected;

β_{scat} is the molecular and aerosol scattering.

3.3.1. Absorption

Absorption is caused by the beam's photons colliding with various finely dispersed liquid and solid particles in the air such as water vapor, dust, ice, and organic molecules. The aerosols that have the most absorption potential at infrared wavelengths include water, O_2, O_3, and CO_2 Absorption has the effect of reducing link margin, distance and the availability of the link [15].

The absorption coefficient depends on the type of gas molecules, and on their concentration. Molecular absorption is a selective phenomenon which results in the spectral transmission of the atmosphere presenting transparent zones, called atmospheric transmission windows [11], shown in Fig. 7, which allows specific frequencies of light to pass through it. These windows occur at various wavelengths. The Atmospheric windows due to absorption are created by atmospheric gases, but neither nitrogen nor oxygen, which are two of the most abundant gases, contribute to absorption in the infrared part of the spectrum [7].

It is possible to calculate absorption coefficients from the concentration of the particle and the effective cross section such as [16,17]:

$$\beta_{abs} = \alpha_{abs} N_{abs} \left[\frac{1}{km} \right]$$ (9)

Where:

α_{abs}: is the effective cross section of the absorption particles [km^2].

N_{abs}: is the concentration of the absorption particles [$1/km^3$].

An absorption lines at visible and near infrared wavelengths are narrow and generally well separated. Thus, absorption can generally be neglected at wavelength of interest for free space laser communication. Another reason for ignoring absorption effect is to select wavelengths that fall inside the transmittance windows in the absorption spectrum [18].

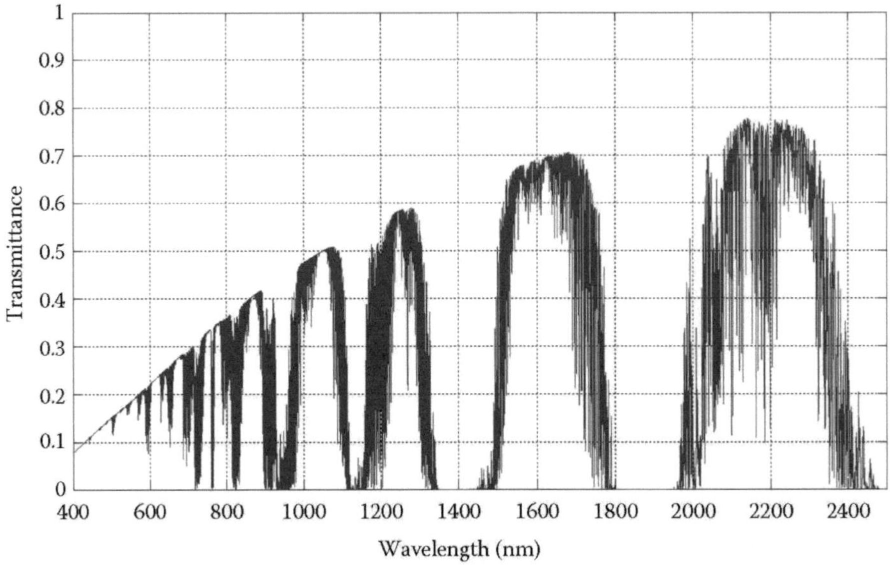

Figure 7. Atmospheric transmittance window with absorption contribution.

3.3.2. Scattering

Scattering is defined as the dispersal of a beam of radiation into a range of directions as a result of physical interactions. When a particle intercepts an electromagnetic wave, part of the wave's energy is removed by the particle and re-radiated into a solid angle centered at it. The scattered light is polarized, and of the same wavelength as the incident wavelength, which means that there is no loss of energy to the particle [10].

There are three main types of scattering: (1) Rayleigh scattering, (2) Mie scattering, and (3) non-selective scattering. Figure 8 illustrates the patterns of Rayleigh, Mie and non-Selective scattering.

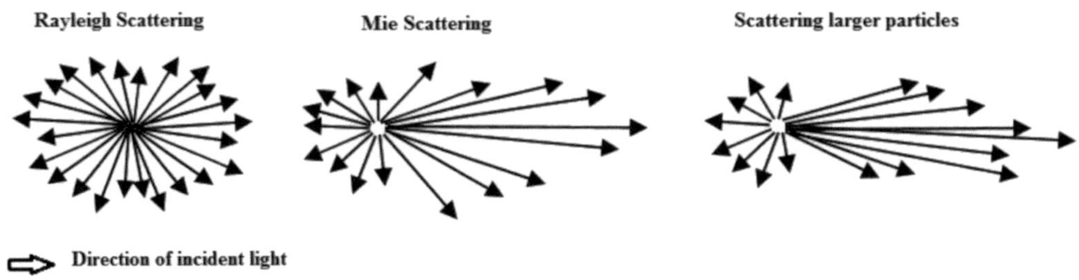

Figure 8. Patterns of Rayleigh, Mie and Non-selective scattering.

The scattering effect depends on the characteristic size parameter x_0, such as that $x_0 = 2\pi r / \lambda$, where, r is the size of the aerosol particle encountered during propagation [19]. If $x_0 << 1$, the backward lobe becomes larger and the side lobes disappear as shown in Fig. 8 [20] and the scattering process is termed as Rayleigh scattering. If $x_0 \approx 1$, the backward lobe is symmetrical with the forward lobe as shown in Fig. 8 and then it is Mie scattering. For $x_0 >> 1$, the particle presents a large forward lobe and small side lobes that start to appear as shown in Fig. 8 [20] and the scattering process is termed as non-selective scattering. The scattering process for different scattering particles present in the atmosphere is summarized in Table (4) [21]. It is possible to calculate the scattering coefficients from the concentration of the particles and the effective cross section such as [16]:

$$\beta_{scat} = \alpha_{scat} N_{scat} [1 / km] \tag{10}$$

Where:

β_{scat}: is either Rayleigh (molecular) β_m or Mie (aerosols) β_a scattering.

α_{scat}: is a cross-section parameters $[km^2]$.

N_{scat}: is a particle concentration $[1 / km^3]$.

The total scattering can be written as:

$$\beta_{scat} = \beta_m + \beta_a [1 / km] \tag{11}$$

Type of particles	Radius (μm)	Size parameter (X_o)	Scattering regime
Air molecules	0.0001	0.00074	Rayleigh
Haze particles	0.01 - 1	0.074 – 7.4	Rayleigh - Mie
Fog droplets	1 - 20	7.4 – 147.8	Mie - Geometrical
Rain droplets	100 - 1000	740 - 7400	Geometrical
Snow flakes	1000 - 5000	7400 - 37000	Geometrical

Table 4. Typical atmospheric scattering parameters, with size parameter.

3.3.2.1. Rayleigh (molecular) scattering

Rayleigh scattering refers to scattering by molecular and atmospheric gases of sizes much less than the incident light wavelength. The Rayleigh scattering coefficient is given by [16]:

$$\beta_m = \alpha_m N_m [1 / km] \tag{12}$$

Where:

α_m: is the Rayleigh scattering cross-section $[km^2]$.

N_m: is the number density of air molecules $[1 / km^3]$.

Rayleigh scattering cross section is inversely proportional to fourth power of the wavelength of incident beam (λ^{-4}) as the following relationship:

$$\alpha_m = \frac{8\pi^3(n^2 - 1)^2}{3N^2\lambda^4}[km^2] \tag{13}$$

Where:

n: is the index of refraction.

λ: is the incident light wavelength $[m]$.

N: is the volumetric density of the molecules $[1 / km^3]$.

The result is that Rayleigh scattering is negligible in the infrared waveband because Rayleigh scattering is primarily significant in the ultraviolet to visible wave range [10].

3.3.2.2. MIE (Aerosol) scattering

Mie scattering occurs when the particle diameter is equal or larger than one-tenth the incident laser beam wavelength, see Table 4. Mie scattering is the main cause of attenuation at laser wavelength of interest for FSO communication at terrestrial altitude. Transmitted optical beams in free space are attenuated most by the fog and haze droplets mainly due to dominance of Mie scattering effect in the wavelength band of interest in FSO (0.5 μm – 2 μm). This makes fog and haze a keys contributor to optical power/irradiance attenuation. The attenuation levels are too high and obviously are not desirable [22].

The attenuation due to Mie scattering can reach values of hundreds of dB/km [19,23] (with the highest contribution arising from fog). The Mie scattering coefficient expressed as follows [10]:

$$\beta_a = \alpha_a N_a [1 / km] \tag{14}$$

Where:

α_a: is the Mie scattering cross-section $[km^2]$.

N_a: is the number density of air particles $[1 / km^3]$.

An aerosol's concentration, composition and dimension distribution vary temporally and spatially varying, so it is difficult to predict attenuation by aerosols. Although their concentration is closely related to the optical visibility, there is no single particle dimension distribution for a given visibility [24]. Due to the fact that the visibility is an easily obtainable parameter, either from airport or weather data, the scattering coefficient β_a can be expressed according to visibility and wavelength by the following expression [11]:

http://dx.doi.org/10.5772/58884

$$\beta_a = \left(\frac{3.91}{V}\right)\left(\frac{0.55\mu}{\lambda}\right)^i \tag{15}$$

Where:

V: is the visibility (Visual Range) $[km]$.

λ: is the incident laser beam wavelength $[\mu m]$.

i: is the size distribution of the scattering particles which typically varies from 0.7 to 1.6 corresponding to visibility conditions from poor to excellent.

Where:

$$i = 1.6 \ \ for \ V > 50 \ km.$$
$$i = 1.3 \ \ for \ 6 \ km \le V \le 50 \ km.$$
$$i = 0.585 \ V^{1/3} \ \ for \ V < 6 \ km.$$

Since we are neglecting the absorption attenuation at wavelength of interest and Rayleigh scattering at terrestrial altitude and according to Eq. 8 and Eq. 11 then:

$$\beta_{scat} = \beta_a \tag{16}$$

The atmospheric attenuation τ is given as:

$$\tau = \exp\left(-\beta_a L\right) \tag{17}$$

The atmospheric attenuation in dB, τ can be calculated as follows:

$$\tau = 4.3429 \beta_a L \ [dB] \tag{18}$$

3.3.2.3. Rain

Rain is formed by water vapor contained in the atmosphere. It consists of water droplets whose form and number are variable in time and space. Their form depends on their size: they are considered as spheres until a radius of 1 mm and beyond that as oblate spheroids: flattened ellipsoids of revolution [11].

Rainfall effects on FSO systems:

Scattering due to rainfall is called non-selective scattering, this is because the radius of raindrops (100 – 1000 μm) is significantly larger than the wavelength of typical FSO systems. The laser is able to pass through the raindrop particle, with less scattering effect occurring. The haze particles are very small and stay longer in the atmosphere, but the rain particles are very large and stay shorter in the atmosphere. This is the primary reason that attenuation via rain is less than haze [24]. An interesting point to note is that RF wireless technologies that use frequencies above approximately 10 GHz are adversely impacted by rain and little impacted by fog. This is because of the closer match of RF wavelengths to the radius of raindrops, both being larger than the moisture droplets in fog [14]. The rain scattering coefficient can be calculated using Stroke Law [25]:

$$\beta_{rain\ scat} = \pi a^2 N_a Q_{scat}\left(\frac{a}{\lambda}\right) \tag{19}$$

Where:

a: is the radius of raindrop, (cm).

N_a: is the rain drop distribution, (cm^{-3}).

Q_{scat}: is the scattering efficiency.

The raindrop distribution N_a can be calculated using equation following:

$$N_a = \frac{R}{1.33(\pi a^3) V_a} \tag{20}$$

Where:

R: is the rainfall rate (cm/s),

V_a: is the limit speed precipitation.

Limiting speed of raindrop [25] is also given as:

$$V_a = \frac{2a^2 \rho g}{9\eta} \tag{21}$$

Where:

ρ: is water density, ($\rho = 1\ g / cm^3$).

g: is gravitational constant, $g = 980\ cm / sec^2$.

η: is viscosity of air, $\eta = 1.8 * 10^{-4} g / cm.sec$.

The rain attenuation can be calculated by using Beer's law as:

$$\tau = \exp\left(-\beta_{rain\ scat} L\right) \tag{22}$$

For more details about several weather conditions and the corresponding visibility at various wavelengths readers can refer to references [26,27].

http://dx.doi.org/10.5772/58884

3.4. Turbulence

Clear air turbulence phenomena affect the propagation of optical beam by both spatial and temporal random fluctuations of refractive index due to temperature, pressure, and wind variations along the optical propagation path [28,29]. Atmospheric turbulence primary causes phase shifts of the propagating optical signals resulting in distortions in the wave front. These distortions, referred to as optical aberrations, also cause intensity distortions, referred to as scintillation. Moisture, aerosols, temperature and pressure changes produce refractive index variations in the air by causing random variations in density. These variations are referred to as eddies and have a lens effect on light passing through them. When a plane wave passes through these eddies, parts of it are refracted randomly causing a distorted wave front with the combined effects of variation of intensity across the wave front and warping of the isophase surface [30]. The refractive index can be described by the following relationship [31]:

$$n - 1 \approx 79 \times \frac{P}{T} \tag{23}$$

Where:

P: is the atmospheric pressure in $[\text{mbar}]$.

T: is the temperature in Kelvin $[\text{K}]$.

If the size of the turbulence eddies are larger than the beam diameter, the whole laser beam bends, as shown in Fig. 9. If the sizes of the turbulence eddies are smaller than the beam diameter and so the laser beam bends, they become distorted as in Fig. 10. Small variations in the arrival time of various components of the beam wave front produce constructive and destructive interference and result in temporal fluctuations in the laser beam intensity at the receiver see Fig. 10.

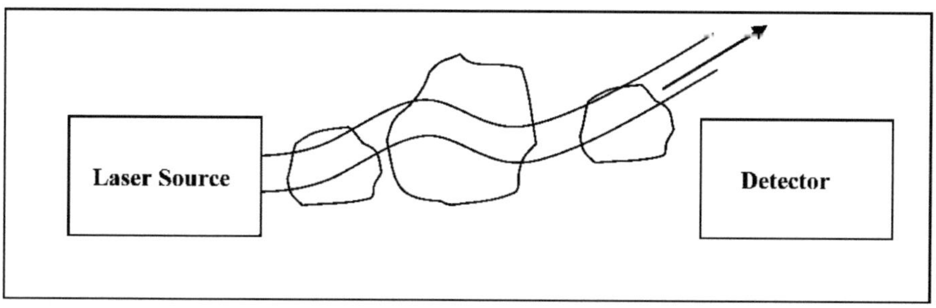

Figure 9. Laser beam Wander Due to turbulence cells that are larger than the beam diameter.

3.4.1. Refractive index structure

Refractive index structure parameter C_n^2 is the most significant parameter that determines the turbulence strength. Clearly, C_n^2 depends on the geographical location, altitude, and time of

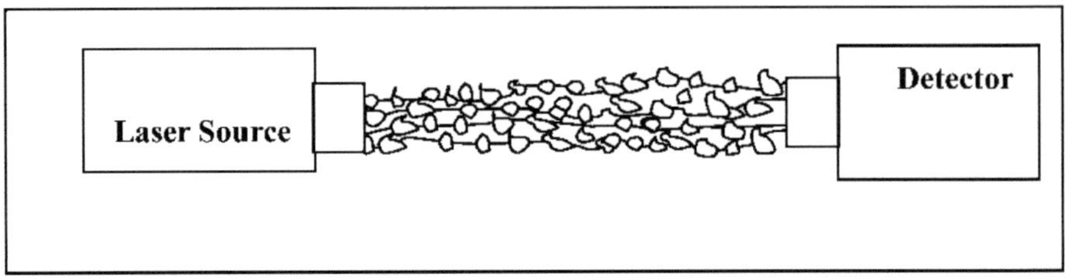

Figure 10. Scintillation or fluctuations in beam intensity at the receiver due to turbulence cells that is smaller than the beam diameter.

day. Close to ground, there is the largest gradient of temperature associated with the largest values of atmospheric pressure (and air density). Therefore, one should expect larger values C_n^2 at sea level. As the altitude increases, the temperature gradient decreases and so the air density with the result of smaller values of C_n^2 [8].

In applications that envision a horizontal path even over a reasonably long distance, one can assume C_n^2 to be practically constant. Typical value of C_n^2 for a weak turbulence at ground level can be as little as $10^{-17}m^{-2/3}$, while for a strong turbulence it can be up to $10^{-13}m^{-2/3}$ or larger. However, a number of parametric models have been formulated to describe the C_n^2 profile and among those, one of the more used models is the Hufnagel-Valley [32] given by:

$$\begin{aligned} C_n^2(h) = 0.00594(v/27)^2(10^{-5}h)^{10} exp\left(-\tfrac{h}{1000}\right) + \\ 2.7\times10^{-16} exp\left(-\tfrac{h}{1500}\right) + A_o exp\left(-\tfrac{h}{100}\right) \end{aligned} \tag{24}$$

Where:

h: is the altitude in $[\mathrm{m}]$.

v: is the wind speed at high altitude $[\mathrm{m/s}]$.

A_0: is the turbulence strength at the ground level, $A_o = 1.7\times10^{-14}m^{-2/3}$.

The most important variable in its change is the wind and altitude. Turbulence has three main effects ; scintillation, beam wander and beam spreading.

3.4.2. Scintillation

Scintillation may be the most noticeable one for FSO systems. Light traveling through scintillation will experience intensity fluctuations, even over relatively short propagation paths. The scintillation index, σ_i^2 describes such intensity fluctuation as the normalized variance of the intensity fluctuations given by [8,14]:

$$\sigma_i^2 = \frac{\langle (I - \langle I \rangle)^2 \rangle}{\langle I \rangle^2} = \frac{\langle I^2 \rangle}{\langle I \rangle^2} - 1 \tag{25}$$

Where:

$I = |E|^2$: is the signal irradiance (or intensity).

The strength of scintillation can be measured in terms of the variance of the beam amplitude or irradiance σ_i given by the following:

$$\sigma_i^2 = 1.23 C_n^2 k^{7/6} L^{11/6} \tag{26}$$

Where C_n^2 is the refractive index structure, $k = 2\pi / \lambda$ is the wave number (an expression suggests that longer wavelengths experience a smaller variance), and l is the link range (m).

Where the Eq. 26 is valid for the condition of weak turbulence mathematically corresponding to $\sigma_i^2 < 1$. Expressions of lognormal field amplitude variance depend on: the nature of the electromagnetic wave traveling in the turbulence and on the link geometry [8].

3.4.3. Beam spreading

Beam spreading describes the broadening of the beam size at a target beyond the expected limit due to diffraction as the beam propagates in the turbulent atmosphere. Here, we describe the case of beam spreading for a Gaussian beam, at a distance l from the source, when the turbulence is present. Then one can write the irradiance of the beam averaged in time as [33]:

$$I(l, r) = \frac{2P_o}{\pi \omega_{eff}^2(l)} exp\left(\frac{-2r^2}{\omega_{eff}^2(l)}\right) \tag{27}$$

Where:

P_o: is total beam power in W

r: is the radial distance from the beam center

The beam will experience a degradation in quality with a consequence that the average beam waist in time will be $\omega_{eff}(l) > \omega(l)$. To quantify the amount of beam spreading, describes the effective beam waist average as:

$$\omega_{eff}(l)^2 = \omega(l)^2 (1 + T) \tag{28}$$

Where:

$\omega(l)$: is the beam waist that after propagation distance L is given by:

$$\omega(l)^2 = \left[\omega_o^2 + \left(\frac{2L}{k\omega_o}\right)^2\right] (m^2) \tag{29}$$

In which ω_o is the initial beam waist at $L = 0$, T: is the additional spreading of the beam caused by the turbulence. As seen in other turbulence figure of merits, T depends on the strength of turbulence and beam path. Particularly, T for horizontal path, one gets:

$$T = 1.33\sigma_i^2 \Lambda^{5/6} \tag{30}$$

While the parameter Λ is given by:

$$\Lambda = \frac{2L}{k\omega^2(l)} \tag{31}$$

The effective waist, $\omega_{eff}(l)$, describes the variation of the beam irradiance averaged over long term.

As seen in other turbulence figure of merits, $\omega_{eff}(l)^2$ depends on the turbulence strength and beam path. Evidently, due to the fact that $\omega_{eff}(l) > \omega(l)$ beam will experience a loss that at beam center will be equal:

$$L_{BE} = 20\ log_{10}(\omega(l) / \omega_{eff}(l)) \tag{32}$$

3.5. Geometric Losses (GL)

The geometric path loss for an FSO link depends on the beam-width of the optical transmitter θ, its path length L and the area of the receiver aperture A_r. The transmitter power, P_t is spread over an area of $\pi(L\theta)^2 / 4$. The geometric path loss for an FSO link depends on the beam-width of the optical transmitter θ, its path length L and the area of the receiver aperture A_r. The transmitter power, P_r is spread over an area of $\pi(L\theta)^{2/4}$. Geometric loss is the ratio of the surface area of the receiver aperture to the surface area of the transmitter beam at the receiver. Since the transmit beams spread constantly with increasing range at a rate determined by the divergence, geometric loss depends primarily on the divergence as well as the range and can be determined by the formula stated as [2]:

$$\text{geometric loss} = \frac{d_2^2}{\left[d_1 + (L\theta)\right]^2} \tag{33}$$

Where:

d_2 is the diameter receiver aperture (unit: m);

d_1 is the diameter transmitter aperture (unit: m);

θ is the beam divergence (unit: mrad);

L is the link range (unit: m).

Geometric path loss is present for all FSO links and must always be taken into consideration in the planning of any link. This loss is a fixed value for a specific FSO deployment scenario; it does not vary with time, unlike the loss due to rain attenuation, fog, haze or scintillation.

3.6. Total attenuation

Atmospheric attenuation of FSO system is typically dominated by haze, fog and is also dependent on rain. The total attenuation is a combination of atmospheric attenuation in the atmosphere and geometric loss. Total attenuation for FSO system is actually very simple at a high level (leaving out optical efficiencies, detector noises, etc.). The total attenuation is given by the following [34]:

$$\frac{P_r}{P_t} = \frac{d_2^2}{\left[d_1 + (L\theta) \right]} \times \exp(-\beta L\ ), \tag{34}$$

where,

P_t is the transmitted power (unit: mW);

P_r is the received power (unit: mW);

θ is the beam divergence (unit: mrad);

β is the total scattering coefficient (unit: km^{-1}).

According to Eq. (34), the variables which can be controlled are the aperture size, the beam divergence and the link range. The scattering coefficient is uncontrollable in an outdoor environment. In real atmospheric situations, for availabilities at 99.9% or better, the system designer can choose to use huge transmitter laser powers, design large receiver apertures, design small transmitter apertures and employ small beam divergence. Another variable that can control is link range, which must be of a short distance to ensure that the atmospheric attenuation is not dominant in the total attenuation [35].

The strength of scintillation can be measured in terms of the variance of the beam amplitude or irradiance σ_i given by the following:

$$\sigma_1^2 = 1.23 C_n^2 k^{7/6} L^{\ 11/6} \tag{35}$$

Here, $k=2\pi/\lambda$ is the wave number and this expression suggests that longer wavelengths experience a smaller variance, and C_n^2 is a refractive index structure parameter. Equation (35) is valid for the condition of weak turbulence mathematically corresponding to $\sigma_i 2<1$. Expressions of lognormal field amplitude variance depend on the nature of the electromagnetic wave traveling in the turbulence and on the link geometry.

In this chapter, we do not take into account the atmospheric turbulence, because its influence in Yemeni climate could be negligible. That means the effect of the turbulence is too small contrary to visibility and geometric loss. Therefore, we have taken into account only the total attenuation depending on visibility, and geometric loss.

An FSO communication system is influenced by atmospheric attenuation, which limits their performance and reliability. The atmospheric attenuated by fog, haze, rainfall, and scintillation has a harmful effect on FSO system. The majority of the scattering occurred on the laser beam is Mie scattering. This scattering is due to the fog and haze aerosols existed at the atmosphere and can be calculated through visibility. FSO attenuation at thick fog can reach values of hundreds dB. Thick fog reduces the visibility range to less than 50 m, and it can affect on the performance of FSO link for distances. The rain scattering (non-selective scattering) is independent on wavelength, and it does not introduce significant attenuation in wireless infrared links, it affects mainly on microwave and radio systems that transmit energy at longer wavelengths. There are three effects on turbulence: scintillation, laser beam spreading and laser beam wander. Scintillation is due to variation in the refractive index of air. If the light is traveled by scintillation, it will experience intensity fluctuations. The geometric loss depends on FSO components design such as beam divergence, aperture diameter of both transmitter and receiver. The total attenuation depends on atmospheric attenuation and geometric loss. To reduce total attenuation, the effect of geometric loss and atmospheric attenuation is small, as FSO system must be designed. The following section explores the simulation results of geometric loss and total attenuations for Yemeni climate.

3.7. Optical link budget

To calculate the FSO link budget several parameters taken into account as geometric loss, link margin, received power and bit error rate. The received power should be grater less the transmitted power from the source and equal the transmitted power minus total loss. In the basic free-space channel the optical field generated at the transmitter propagates only with

an associated beam spreading loss. For this system the performance can be determined directly from the power flow. The signal power received P_{Rx} [W] depends on the transmit power P_{Tx} [W], transmit and receive antenna gains G_{Tx}, G_{Rx}, and the total loss [36]

$$P_{r=} P_t + G_{Tx} + G_{Rx} \text{ - total loss} \tag{36}$$

Table (5) shows the values of transmitted output power for diffuse and tracked topology (Data obtained from [7]).

In the indicated reference, they presented an expression to calculate the link distance L achievable from direct line propagation:

Data Bit Rate (Mbps)	Optical Power (Diffuse System)	Optical Power (Tracked System)	Optical Power (Cellular System)
0.1	$3 \times 10^{-2}\ W$	$6 \times 10^{-7}\ W$	$3 \times 10^{-3}\ W$
0.5	$6 \times 10^{-2}\ W$	$8 \times 10^{-7}\ W$	$5.8 \times 10^{-3}\ W$
1	$8 \times 10^{-2}\ W$	$1 \times 10^{-6}\ W$	$8 \times 10^{-3}\ W$
5	$1 \times 10^{-1}\ W$	$3 \times 10^{-6}\ W$	$1 \times 10^{-2}\ W$
10	$2.5 \times 10^{-1}\ W$	$6 \times 10^{-6}\ W$	$2 \times 10^{-2}\ W$
50	$6 \times 10^{-1}\ W$	$9 \times 10^{-6}\ W$	$3 \times 10^{-2}\ W$

Table 5. Values of transmitted output power for diffuse and tracked topology from [7].

$$L = \frac{P_t A_r T_1 T_2}{2\pi P_{rm}(1 - \cos \emptyset)} \tag{37}$$

Here, P_t represent the optical output power from the transmitter (in mW), A_r is the active area of the photodetector, T_1 is the transmittivity of the transmitter filter, T_2 is the transmittivity of the filter at the receiver, P_{rm} is the optical power required (in mW) to obtain a specific carrier-to-noise ratio at the receiver, and $\emptyset$ is the half angle of the energy related by optical source. From this expression, they calculate achievable distances (depending on the FOV), which in their case covered a range of between 10 and 20 m.

3.8. BER and SNR

Both SNR and BER are used to assess the quality of communication systems. BER performance depends on the average received power, the scintillation strength, and the receiver noise. With an appropriate design of aperture averaging, the received optical power could be increased and the effect of the scintillation can be dumped. With turbulence, the SNR is expressed as follows [37]:

$$SNR = \left(0.31 C_n^2 k^{\frac{7}{6}}\, l^{\frac{11}{6}}\right)^{-1} \tag{38}$$

For FSO links with an on off keying modulation scheme in BER can be written as

$$BER = \frac{exp(-SNR / 2)}{(2\pi SNR)^{0.5}} \tag{39}$$

In our model, we have assumed that the surface area of the photo detector is large enough so that the effective SNR includes the beam spreading effect, thus the effective SNR is defined as

$$SNR_{eff} = \frac{SNR}{1 + 1.33\sigma_i^2 \left[\frac{2l}{k\omega(l)^2} \right]^{5/6}} \tag{40}$$

The performance and reliability of FSO communication systems are affected and limited by atmospheric attenuation. It has a harmful effect by haze, rainfall, fog, and scintillation has a harmful effect of FSO system. The majority of the scattering occurred to the laser beam is due to the Mie scattering. This scattering is due to the fog and haze aerosols existed at the atmosphere. This scattering is calculated through visibility. FSO attenuation at thick fog can reach values of hundreds dB. Thick fog reduces the visibility range to less than 50 m, and it can affect on the performance of FSO link for distances as small. The rain scattering (non-selective scattering) is wavelength independent and it does not introduce a significant attenuation in wireless IR links, it affect mainly on microwave and radio systems that transmit energy at longer wavelengths.

There are three effects on turbulence: scintillation, laser beam spreading and laser beam wander. Scintillation is due to variation in the refractive index structure of air, so if the light traveling through scintillation, it will experience intensity fluctuations. The Geometric loss depends on FSO components design such as beam divergence, aperture diameter of both transmitter and receiver. The total attenuation depends on atmospheric attenuation and Geometric loss. In order to reduce total attenuation, FSO system must be designed so that the effect of geometric loss and atmospheric attenuation is small.

4. Practical part: Case study

In this chapter, we will take Yemeni climate as a case study to study and analyze the practical part of FSO system by series of simulations obtained results.

4.1. Simulation results and discussion of geometric loss and total attenuation

4.1.1. Geometric loss

This part illustrates the effects of geometric loss on the performance of FSO system. We calculated the value of geometric loss using Eq. (33) assuming that the link range is 1 km and beam divergence is 1 mrad at two different designs, which are considered as particular design specifications shown in Table 6, due to particular implementation especially based on the existing product available in the industry [38,39].

design	diameter of transmitter aperture	diameter of receiver aperture
design 1	8 cm	10 cm
design 2	3.5 cm	7 cm

Table 6. Diameters of transmitter and receiver aperture of an FSO system.

http://dx.doi.org/10.5772/58884

There are a number of parameters that control geometric loss: transmission range, the diameter of transmitter and receiver apertures and laser beam divergence. These parameters also contribute to the design of FSO system, so that it is suitable during bad weather conditions.

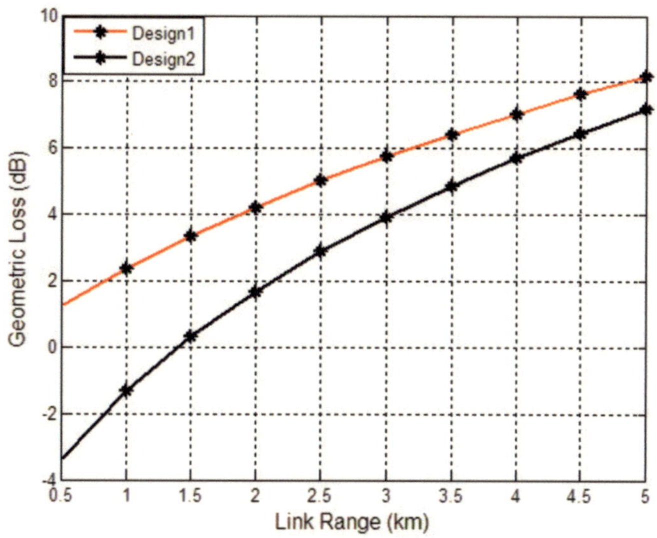

Figure 11. Geometric loss (dB) versus link range (km).

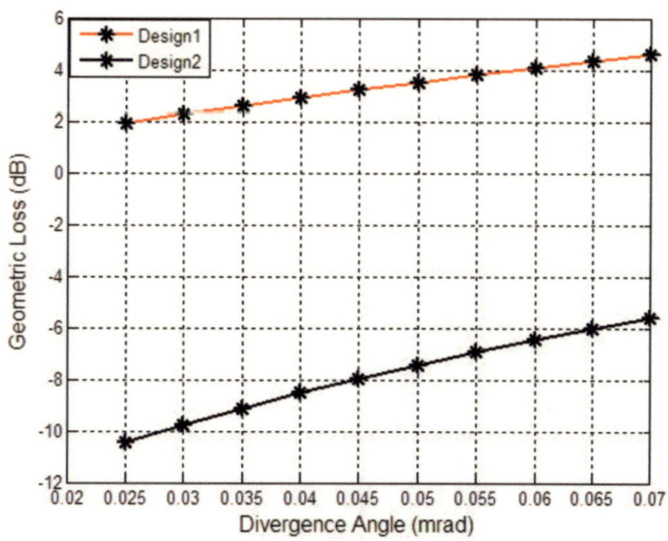

Figure 12. Geometric loss (dB) versus divergence angle (mrad).

Figure 11 shows the geometric loss versus link range using the values presented in Table 6 and divergence angle is about 0.025 mrad. The link range is in the range of 0.5 to 5.0 km. Geometric loss is proportional to link range, which shows that the link range increases with the increases of geometric loss. As demonstrated in Fig. 11 the geometric loss is 1.3 dB at 0.5 km for design 1 and -3.4 dB for design 2. While at the distance of 5km the geometric loss for design 1 reaches 8.2 dB and 7.2 dB for design 2. Figure 12 illustrates the geometric loss versus the divergence angle. The divergence angle is in the range of 0.025 to 0.07 mrad. Geometric loss is proportional to divergence angle, which suggest that when the divergence angle increases, geometric loss enhances. For a 0.025 mrad divergence angle, the geometric loss is about 1.93 dB for design 1 and -10.5 dB for design 2. For a 0.07 mrad divergence angle, the geometric loss is about 4.6 dB for design 1 and -5.6 dB for design 2. That means by using a small divergence angle of laser beam in FSO systems, geometric loss effect is minimized.

Figure 13 demonstrates the geometric loss versus the transmitter aperture diameter using the values presented in Table 1, divergence angle is about 0.025 mrad and the link range is 1 km. The transmitter aperture diameter is in the range of 2 to 22 cm. This figure shows that the transmitter aperture diameter rises with increases of the geometric loss. For transmitter aperture diameter of 2 cm, the geometric loss is about -7 dB for design 1 and -3.87 dB for design 2. For the transmitter aperture diameter of 20 cm, the geometric loss is about 7.7 dB for design 1 and 10.2 dB for design 2. That means the small transmitter aperture diameter is suggested to minimize in the geometric loss effect on FSO systems.

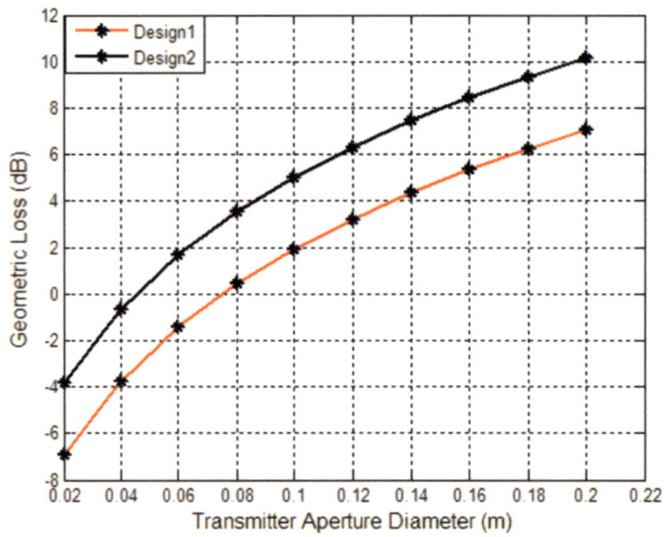

Figure 13. Geometric loss (dB) versus transmitter aperture diameter (m).

Figure 14 indicates the geometric loss versus the receiver aperture diameter using the values presented in Table 7, divergence angle is about 0.025 mrad and the link range is 1 km. When the receiver aperture diameter increases, the geometric loss decreases. For receiver aperture diameter of 2 cm, the geometric loss is about 14.4 dB for design 1 and 9.5 dB for design 2. For the receiver aperture diameter of 20 cm, the geometric loss is about -5.6 dB for design 1 and -10.5 dB for design 2.

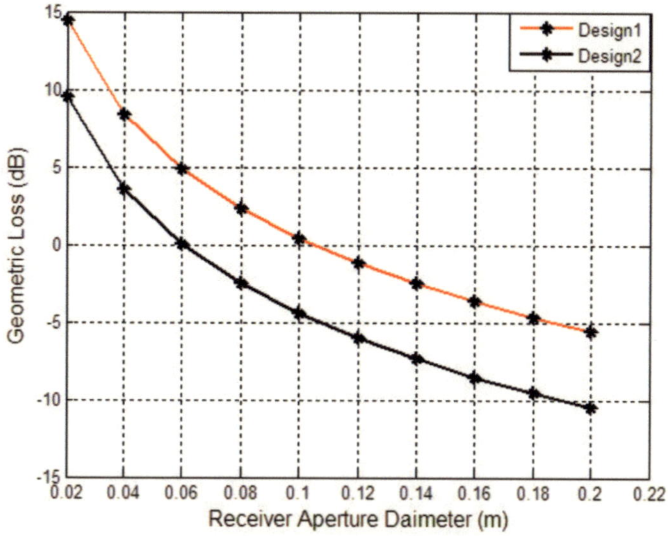

Figure 14. Geometric loss (dB) versus receiver aperture diameter (m).

That is to say that the large receiver aperture diameter will be used to reduce the geometric loss effect on the FSO systems. The results of geometric losses with design parameters are presented in Table 7. We note that the geometric loss at low values for receiver aperture diameter is high compared to the upper values. Because the aperture diameter of receiver is smaller than aperture diameter of transmitter. At a result, the aperture diameter of transmitter must be smaller than at the receiver side.

design parameters	geometric loss/dB			
	design 1		design 2	
	from	to	from	to
link range	1.3	8.2	-3.4	7.2
divergence angle	1.93	4.6	-10.5	-5.6
receiver aperture diameter	14.8	-5.2	10.4	-9.8
transmitter aperture diameter	-6	7.2	-2.9	10.3

Table 7. Results of geometric loss with design parameters.

4.1.2. Total attenuation

Total attenuation depends on attenuation resulted from hazy and rainy days and geometric loss. The attenuation in hazy days depends on visibility, while during rainy days it would be determined by rainfall rate. Visibility range changes with the quantity and density of particles, such as fog, haze and dust attached to air. The higher the density of these particles is, the less visibility is and total attenuation increases. The density of these particles is not fixed. It keeps varying with time and place as well as rainfall. The quantity and density of these particles are unpredictable, and visibility and rainfall rate are also uncontrollable. Thus, they are all not part of FSO design.

We can control the value of geometric loss, because it depends on fixed parameters such as transmitter diameter and receiver apertures, transmission range, and beam divergence. During the design of FSO system, geometric loss must be at minimum to reduce the effect of total attenuation on FSO system. In this part, we used design 2 as demonstrated in Table 6 to calculate total attenuation because this design geometric loss is less as described above in Section 4.1.1.

4.1.2.1. Total attenuation during hazy days

Figure 15 shows total attenuation at low visibility. We used Eq. (8) to plot Fig. 15. Here, we assume that link range is 1 km. When visibility is 0.8 km, total attenuations are 31.8, 31 and 26.4 dB at wavelengths of 780, 850 and 1550 nm, respectively. And when visibility is 5 km, total attenuation are 16.8, 16.4, and 15.4 dB for wavelengths of 780, 850 and 1550 nm, respectively. It is suggested that the more visibility is the least effects of total attenuations on FSO performance.

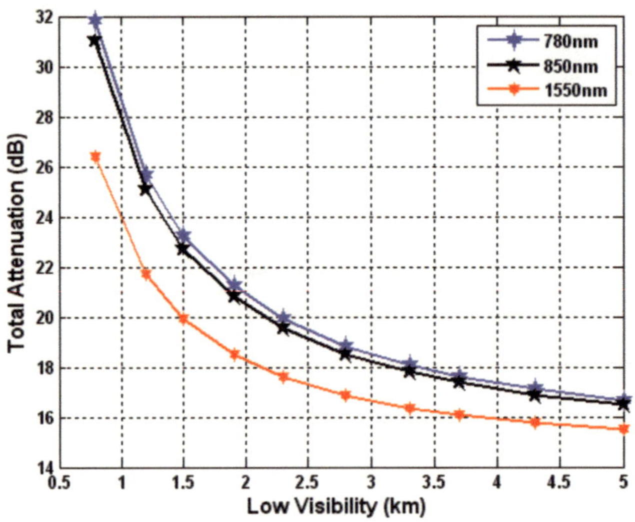

Figure 15. Total attenuation (dB) versus low visibility (km).

Figure 16 indicates the total attenuation versus average visibility. When visibility is 6.4 km, total attenuation is 15.96, 15.8 and 14.9 dB at wavelengths of 780, 850 and 1550 nm, respectively. Note that when visibility is 9.7 km, total attenuation are of 15.39, 15.28 and 14.7 dB at wavelengths of 780, 850 and 1550 nm, respectively. Based on the previously mentioned, we conclude that total attenuation at wavelength 1550 nm is less than that at wavelengths of 780 and 850 nm. Therefore, to reduce the effect of total attenuation during hazy days, we use the wavelength of 1550 nm.

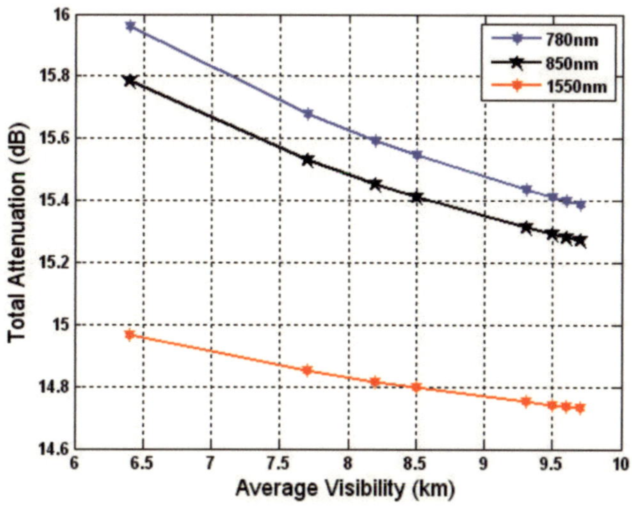

Figure 16. Total attenuation (dB) versus average visibility (km).

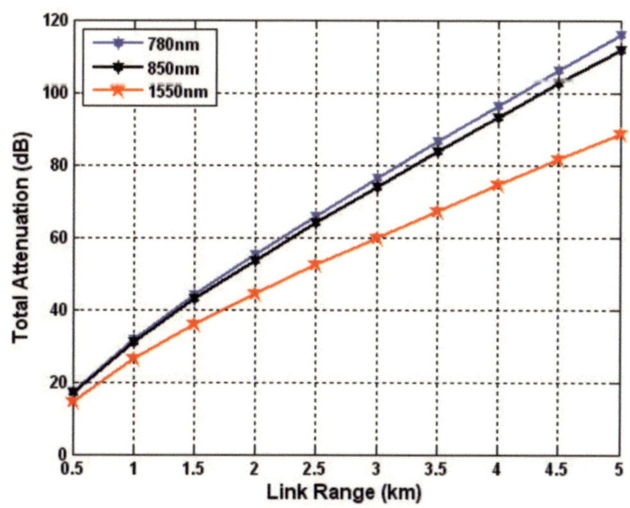

Figure 17. Total attenuation (dB) versus link range (km).

Figure 17 represents total attenuation versus link range. From this figure, we found that total attenuation directly proportions with link range. When link range is 0.5 km, total attenuation is 17.3, 16.9 and 14.6 dB at wavelengths of 780, 850 and 1550 nm, respectively. In addition, when link range is 5 km, total attenuation becomes 115.0, 111.8 and 88.4 dB at wavelengths of 780, 850 and 1550 nm, respectively. Therefore, to reduce the effect of total attenuation on FSO, the distance between the transmitter and receiver shall be small. Figure 18 shows the relationship between total attenuation and laser beam divergence for three wavelengths. With increasing the beam divergence, the total attenuations are increased for three cases as demonstrated in Fig. 18.

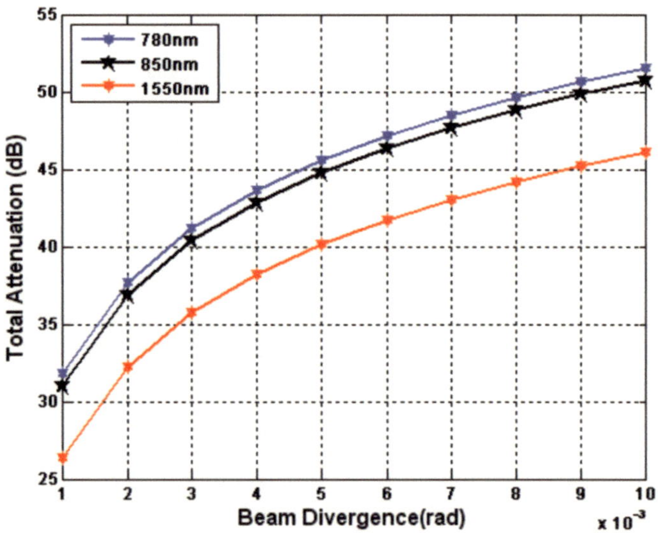

Figure 18. Total attenuation (dB) versus laser beam divergence (mrad).

That means when the beam divergence at 1 mrad the total attenuations 32, 31, and 26 for wavelengths 780, 850, and 1550 nm, respectively. While at beam divergence of 10 mrad, we noticed that the total attenuation was increased 51.1, 50.8, and 46 dB for three previously indicated wavelengths. Therefore, to reduce atmospheric attenuation, the beam divergence should be small in accordance with the previous results. Table 8 shows the results of total attenuation for design parameters at hazy days.

4.1.2.2. Total attenuation in rainy days

Figure 19 shows the total attenuation versus rainfall rate. It can be seen obviously that the influence of attenuation on transmission of FSO systems is more prominent during heavy rainfall compared to moderate and light rainfall. Figure 20 indicates the total attenuation versus link range. The atmospheric attenuation is proportional to link range, which showed

http://dx.doi.org/10.5772/58884

parameters	wavelength	total attenuation/dB	
		from	to
low visibility	780 nm	31.8	16.8
	850 nm	31	16.4
	1550 nm	26.4	15.4
average visibility	780 nm	15.96	15.3
	850 nm	15.8	15.3
	1550 nm	14.9	14.7
link range	780 nm	17.3	115
	850 nm	16.9	111.8
	1550 nm	14.6	88.4
beam divergence	780 nm	32	51.6
	850 nm	31	50.8
	1550 nm	26	46

Table 8. Results of total attenuation for design parameters at hazy days.

that when the link range increases, the total attenuation would increase as well. The results of total attenuation for design parameters at rainy days are presented in Table 9.

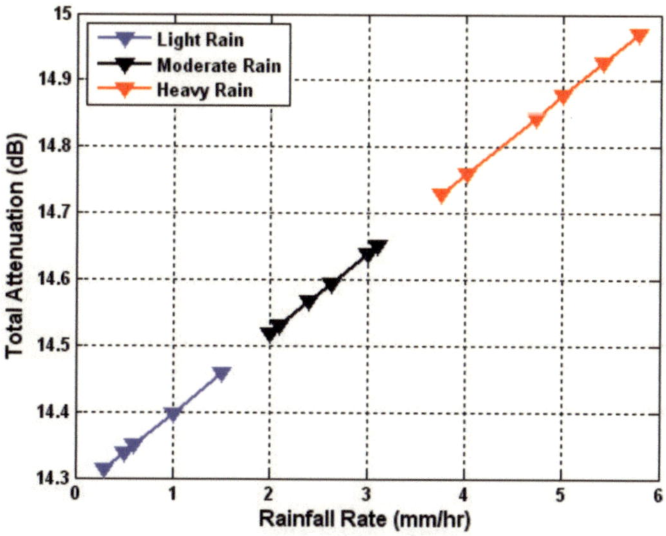

Figure 19. Total attenuation (dB) versus rainfall rate (mm/hr).

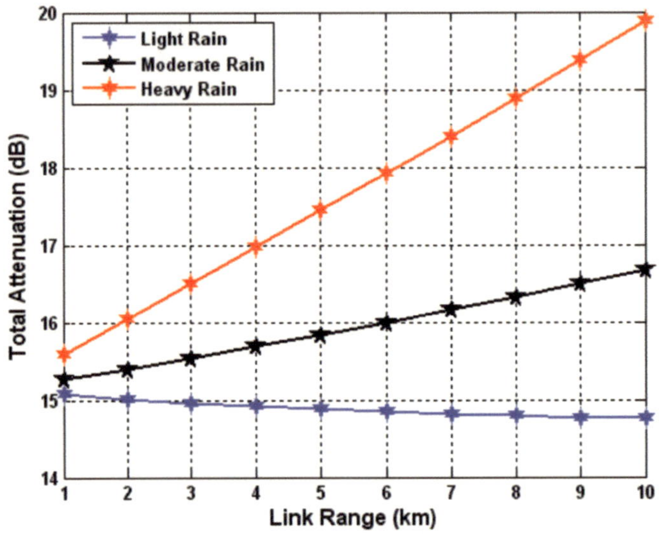

Figure 20. Total attenuation (dB) versus link range (km).

	rainfall	total attenuation/dB	
		from	to
rainfall rate	light	14.3	14.46
	moderate	14.5	14.67
	heavy	14.71	14.98
link range	light	15.1	15.5
	moderate	15.4	16.8
	heavy	15.6	20

Table 9. Results of total attenuation for design parameters at rainy days parameters.

4.2. Simulation results and discussion of haze effects on FSO in Sana'a, Aden, and Taiz cities

In this section, we study the effect of atmospheric attenuation and scattering coefficient on the performance of FSO system in environment of Sana'a, Aden and Taiz cities.

4.2.1. Sana'a city

Low visibility range for Sana'a city in Fig. 21 extends from 0.3 to 6 km. Scattering coefficient at low visibility of 0.3 km is 11.37, 10.99 and 8.69 km^{-1} for wavelengths of 780, 850 and 1550 nm, respectively. The scattering coefficient of 6 km low visibility is 0.45, 0.41 and 0.21 km^{-1} for wavelengths of 780, 850, and 1550, respectively.

http://dx.doi.org/10.5772/58884

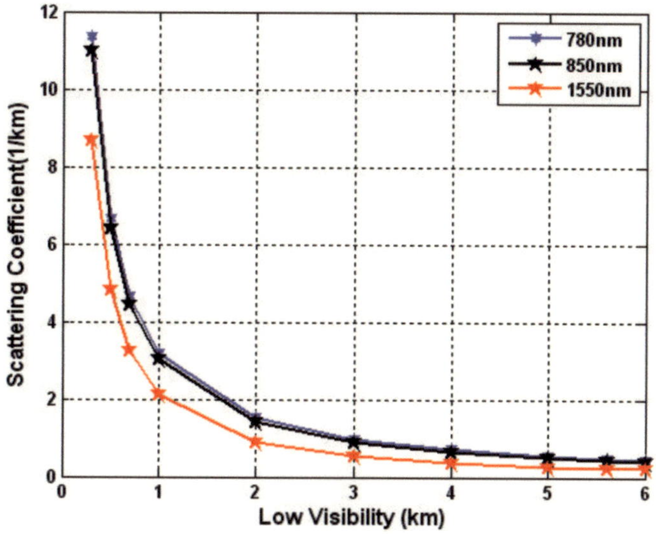

Figure 21. Scattering coefficient (km^{-1}) versus low Visibility for Sana'a city (km).

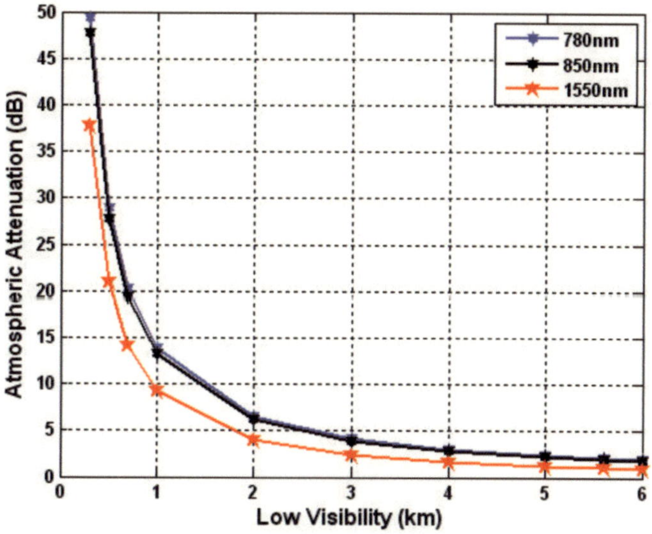

Figure 22. Atmospheric attenuation (dB) versus low visibility (km) for Sana'a City.

Figure 22 shows that the atmospheric attenuation versus low visibility in Sana'a city. At low visibility of 0.3 km, atmospheric attenuation is 49.4 dB, 47.7 dB and 37.7 dB for wavelengths of 780, 850 and 1550 nm, respectively. For 6 km low visibility, the atmospheric attenuation is about 2 dB, 1.8 dB and 0.94 dB for wavelengths of 780, 850 and 1550 nm, respectively.

4.2.2. *Aden city*

Low visibility range in Fig. 23 extends from 0.05 to 6 km for Aden city. Scattering coefficient at low visibility 0.05 km is 44.8, 43.8 and 37.6 km^{-1} for wavelengths of 780, 850 and 1550 nm, respectively. The scattering coefficient of 6 km low visibility is 0.45, 0.41 and 0.22 km^{-1} for wavelengths 780, 850, and 1550 nm, respectively.

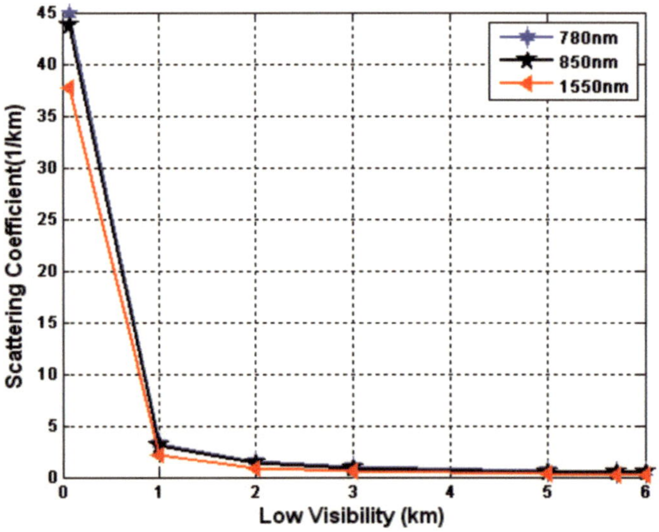

Figure 23. Scattering coefficient (km^{-1}) versus low visibility (km) for Aden city.

Figure 24 shows that the atmospheric attenuation versus low visibility for Aden city. At low visibility of 0.05 km, atmospheric attenuation is 194.4, 190.2 and 163.5 dB for wavelengths 780, 850 and 1550 nm, respectively. For 6 km low visibility, the atmospheric attenuation is about 1.95, 1.8 and 0.94 dB for wavelengths of 780, 850 and 1550 nm, respectively.

4.2.3. *Taiz city*

Low visibility range in Fig. 25 extends from 0.05 to 4 km for Taiz city. Scattering coefficient at low visibility 0.05 km is 44.8, 43.8 and 37.6 km^{-1} for wavelengths of 780, 850 and 1550 nm, respectively. The scattering coefficient of 4 km low visibility is 0.7, 0.65 and 0.37 km^{-1} for wavelengths of 780, 850, and 1550 nm, respectively. Figure 26 shows that the atmospheric attenuation versus low visibility for Taiz city. At low visibility of 0.05 km, atmospheric attenuation is 194.4, 190.2 and 163.5 dB for wavelengths of 780, 850 and 1550 nm, respectively. For 4 km low visibility, the atmospheric attenuation is about 3.1, 2.8 and 1.6 dB for wavelengths of 780, 850 and 1550 nm, respectively. Table 10 shows the results of scattering coefficient and atmospheric attenuation at low visibility for Sana'a, Aden and Taiz Cities. The results show

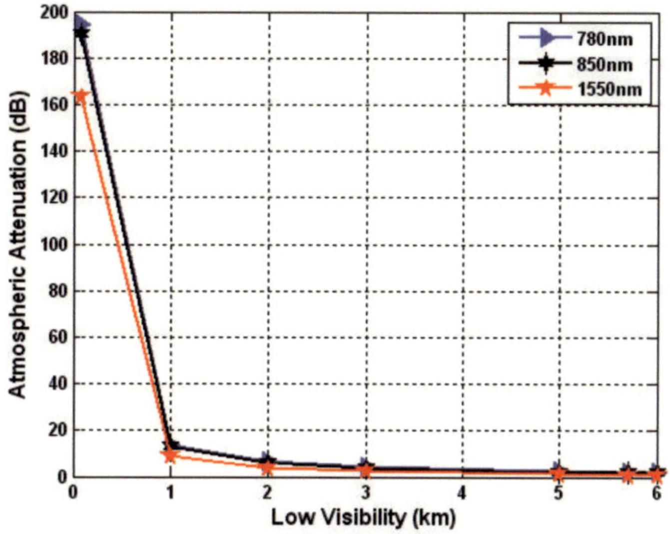

Figure 24. Atmospheric attenuation (dB) versus low visibility (km) for Aden city.

that with increasing the wavelength, in consequence the scattering coefficient and atmospheric attenuation decreased for the three cases were studied in this paper.

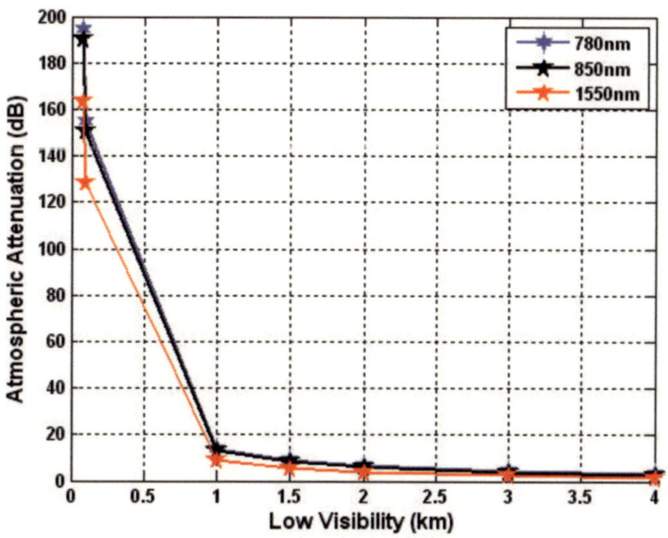

Figure 25. Scattering coefficient (km^{-1}) versus low versus low visibility (km) for Taiz city.

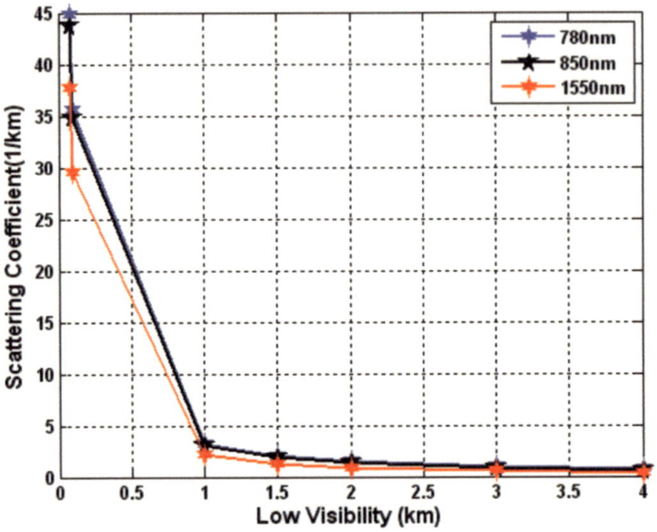

Figure 26. Atmospheric attenuation (dB) versus low visibility (km) for Taiz city.

city	wavelength	scattering coefficient/km^{-1}		atmospheric attenuation/dB	
		from	to	from	to
Sana'a	780 nm	11.37	0.45	49.4	2
	850 nm	10.99	0.41	47.7	1.8
	1550 nm	8.69	0.21	37.7	0.94
Aden	780 nm	44. 8	0.45	194.4	1.95
	850 nm	43.8	0.41	190.2	1.8
	1550 nm	37.6	0.22	163.5	0.94
Taiz	780 nm	44.8	0.7	194.4	3.1
	850 nm	43.8	0.65	190.2	2.8
	1550 nm	37.6	0.37	163.5	1.6

Table 10. Results of scattering coefficient and atmospheric attenuation at low visibility for Sana'a, Aden and Taiz cities.

5. Optical link budget

After illustrating the geometric loss, total attenuation, and haze effects on the FSO in Sana'a, Aden and Taiz cities, we return to the link budget of FSO systems. This section concentrates on received power versus low and average visibility, and link range. Table 11 illustrates the main FSO link parameters. We note that all the given values in the following table are presumed to calculate the received power for three cases as presented in Figs. 27–29.

parameters	description
wavelength (λ)	780, 850, 1550 nm
transmit power (P_{tx})	23.52 dB
beam divergence	1 mrad
visibility	5 km

Table 11. Optical link budget parameters.

The results presented in Fig. 27 show the relationship between received power for three different wavelengths and low visibility. As seen in Fig. 27, received power increases with the increment of low visibility. We note that the obtained received power at the wavelength of 1550 nm is the best as compared to the other two. For example, the received power curve for the wavelength of 1550 nm increases from −67 dBm at the distance of 0.6 km to –27 dBm at the distance of 5 km. However, we note that the receiving power is reduced for other two wavelengths of 780 and 850 nm. As shown in Fig. 28, the received power at wavelength of 1550 nm shows the best compared to other two wavelengths. While the received powers at the wavelengths of 780 and 850 nm are, lower.

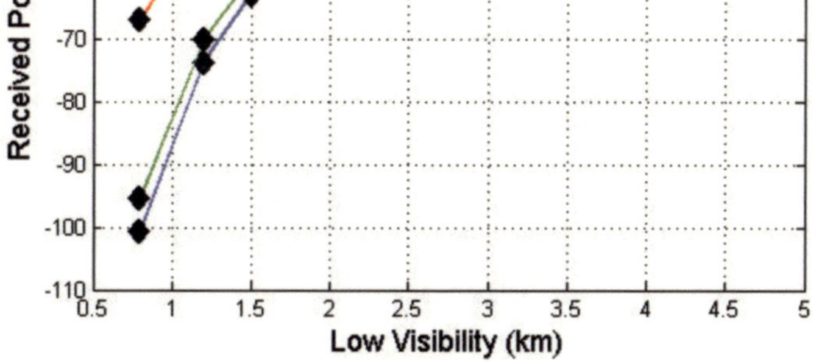

Figure 27. Received power (dBm) versus low visibility (km).

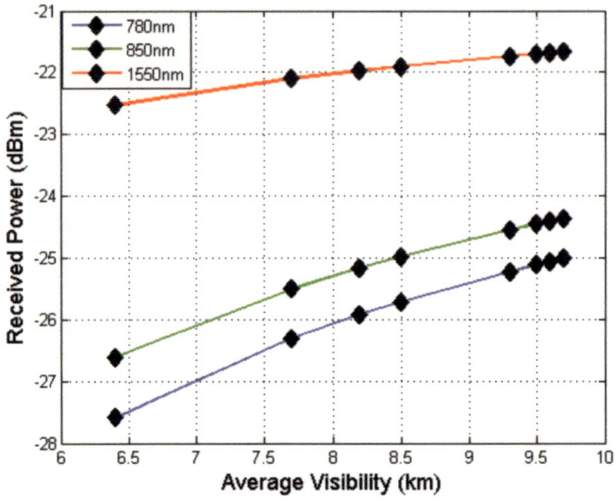

Figure 28. Received power (dBm) versus average visibility (km).

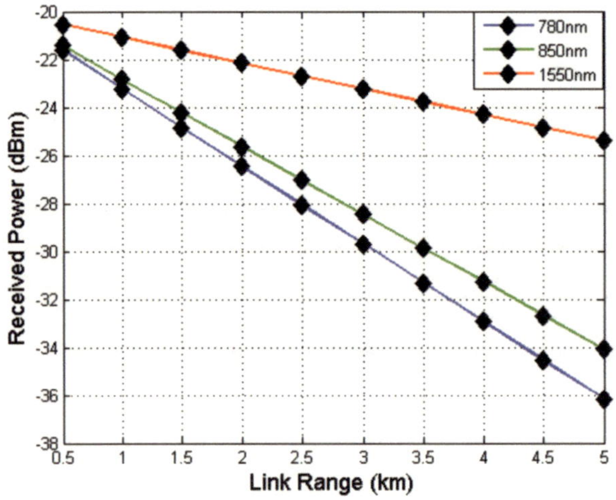

Figure 29. Received power (dBm) versus link range (km).

Figure 29 shows the received power versus the link range. As the link range between transmission and receiver increases, the received power decreases. At the distance of 0.5 km, the received power for the wavelength of 1550 nm is of –20.3 dBm where for the others two are of –21.7 dBm. However, in the distance of 5 km, the received power reaches –36 dBm for wavelength of 780 nm and –34.1 dB for the wavelength of 850 nm. For three study cases, the study was done to improve the efficiency of FSO systems, the wavelength of 1550 nm for three cases must be used and the distance between transmitter and receiver should be reduced.

http://dx.doi.org/10.5772/58884

6. Simulation results of BER and SNR

The data was taken from the Civil Aviation and Meteorology Authority and the Yemeni Meteorological Service. The work includes the analysis of these real data. The purpose here is to discuss the relationship for calculating the variance, SNR, and BER for a range of parameters. We used the wavelengths of 850, 1000, and 1550 nm. Particular attention was given to the 1550 nm wavelength since it is commonly used as the 3rd window of optical communication backbone links. Moreover, being significantly bigger than visible wavelengths, the human retina in particular and the components of the eye in general are less sensitive to the 1550 nm wavelength. Thus, this wavelength is appropriate for eye safety.

Figure 30 illustrates the log intensity fluctuations versus the link range between transmitter and receiver for three values of wavelengths. The log intensity fluctuation depends on the wavelength and increases with the propagation distance. As the transmission range increases the variance (atmospheric turbulence) increases too. For a 2000 m transmission range, the variance is about 0.17 for the wavelength of 850 nm, 0.12 for 1000 nm, and 0.075 for 1550 nm. For a 4000 m transmission range, the variance is about 0.56 for 850 nm, 0.42 for 1000 nm, and 0.25 for 1550 nm. These results show that the use of a wavelength of 1550 nm can reduce the variance "atmospheric turbulence" effect on the FSO systems [37].

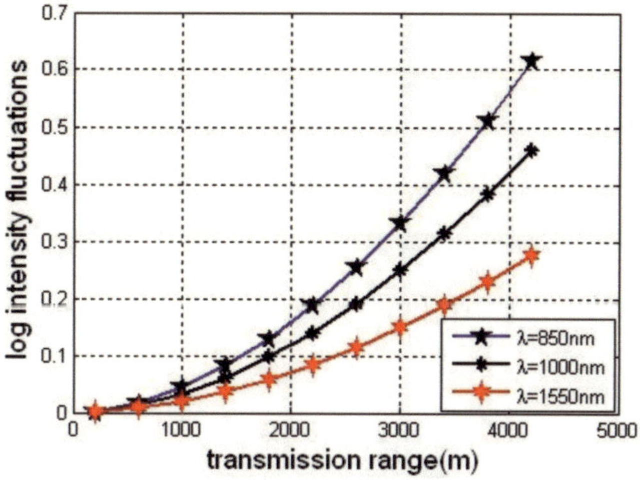

Figure 30. Intensity fluctuations against transmission range.

Figure 31 indicates the comparison between the beam spreading on a distance l from the transmitter in case of atmospheric turbulence and in case without atmospheric turbulence. The spot size of the beam at the transmitter (with the distance $l = 0$) equals 0.008 m. At the distance 200 m, the spot size of the beam is $\omega(l) = 0.015$ m in case of absent turbulence and $\omega_{eff}(l) = 0.015$ m in case of turbulences. At the distance 5000 m, the $\omega(l) = 0.31$ m and $\omega_{eff}(l) = 0.33$ m. From the above results, we conclude that the expansion of the spot size of the beam depends on the distance between sender and receiver as indicated on Fig.32, and on the atmospheric turbulence along the transmission range as indicated on Fig. 33. The higher the turbulence is, the greater the expansion of the beam size is. Figure 34 shows the SNR versus the transmission range of 0 to 4500 m. As the link range between the transmitter and receiver increases, the SNR decreases. This means that the increment of link range is able to decrease the transmission quality and efficiency of FSO systems. At a low range of 200 m, the SNR is about 74 dB for 850 nm, 77 dB for 1000 nm, and 82 dB for 1550 nm. For 4000 m, the SNR is about 18 dB for 850 nm, 21 dB for 1000 nm, and 26 dB for 1550 nm.

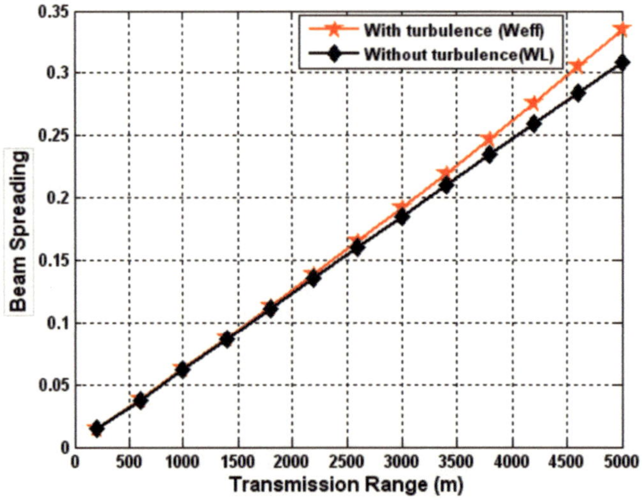

Figure 31. Beam spreading versus transmission range.

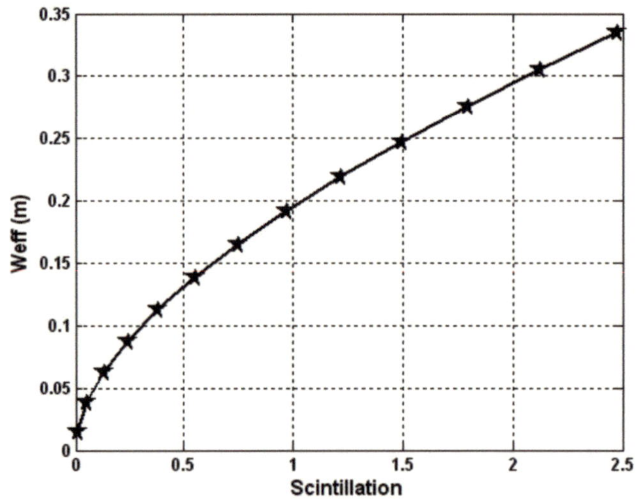

Figure 32. Waist of beam with turbulence versus scintillation.

http://dx.doi.org/10.5772/58884

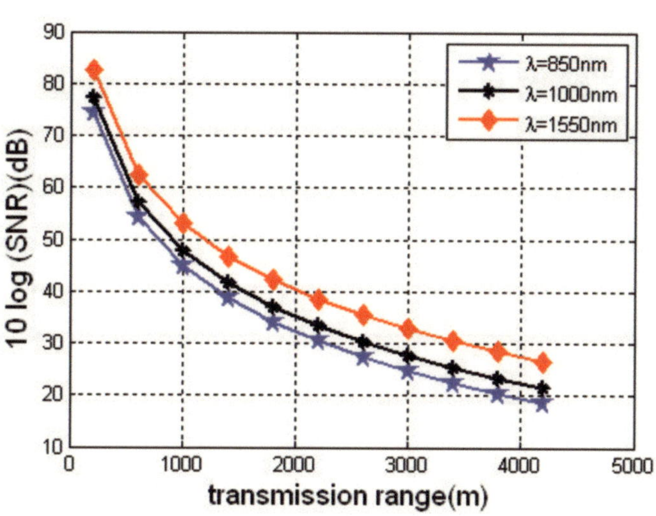

Figure 33. SNR versus transmission range.

Figure 34 shows the BER versus the transmission range. As the link range between transmission and receiver increases, the BER increases too. At 2500 m link range, the BER is about 10^{-4} for 850 nm, 10^{-6} for 1000 nm, and 10^{-9} for 1550 nm. At 4000 m, the BER is 10^{-2} for 850 nm, 10^{-3} for 1000 nm, and 10^{-4} for 1550 nm. If we want an acceptable communication BER of 10^{-9}, the maximum link range between transmitter and receiver should be about 1600 m for 850 nm, 1900 m for 1000 nm, and 2500 m for 1550 nm.

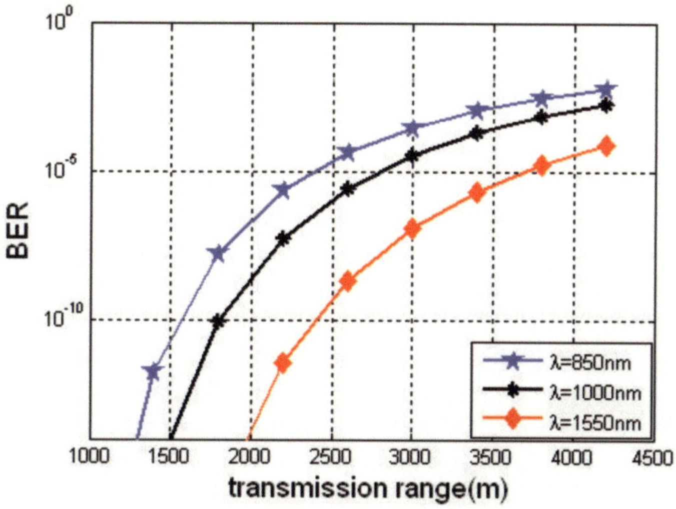

Figure 34. BER versus transmission range.

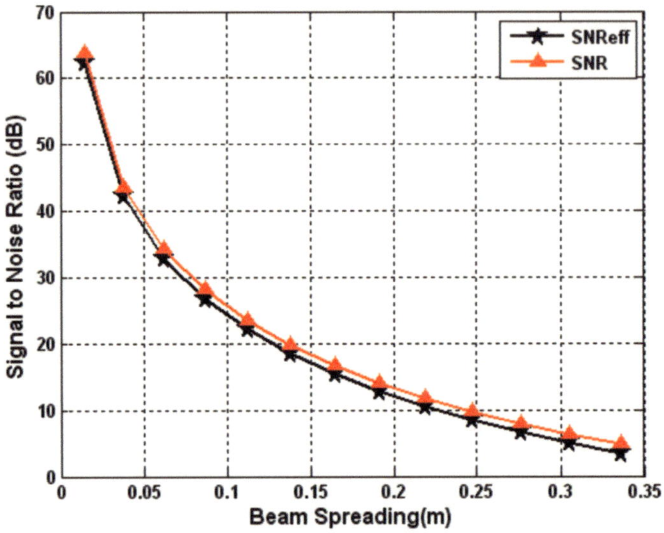

Figure 35. SNR with and without turbulence, SNR_{ref} and SNR, respectively, versus transmission range.

Figure 35 indicates SNR versus expansion of the beam size resulting from air turbulence. For a beam size of ω_{eff} (l) = 0.015 m, the SNR = 64 dB and SNR_{eff} = 62 dB, but for ω_{eff} (l) = 0.33 m, the SNR = 4.7 dB and SNR_{eff} = 3.4 dB. From these results, we conclude that when the beam expands, the loss in terms of the beam intensity increases. This leads to the decrease in the SNR value, and therefore the BER increases as indicated in Fig. 36. For a spot size of 0.015 m, the BER = 10^{-115}, and when the spot size of the beam is 0.33 m, the BER increases up to 10^{-5} approximately. From the results above, we conclude that the narrow beam shows a limited effect of the atmospheric turbulence on the intensity.

Figure 37 shows the BER versus link range between transmitter and receiver. This figure graphically represents the BER as a function of the irradiance variance. For 3500 m, BER is 10^{-6} for the SNR and 10^{-5} for the SNR_{eff}. For an irradiance variance 0.05 the BER 10^{-6} for the SNR and 10^{-5} for the SNR_{eff}. From the results obtained, we conclude that to improve the performance of the FSO transmission systems, it is recommended to shorten the link range between transmitter and receiver. Another improvement of the signal quality offered by the FSO systems includes using the 1550 nm wavelength. The SNR of FSO systems with 1550 nm wavelength is higher than that corresponding to 1000 and 850 nm wavelengths. To reduce the atmospheric turbulence effects on FSO systems, we suggest using the 1550 nm wavelength. Moreover, for the 1550 nm wavelength, the allowable power is largely higher compared to smaller wavelengths (about 50 times compared to 850 nm). This shows that the system operates well during heavier attenuation of the atmosphere since we can safely increase power at the source.

http://dx.doi.org/10.5772/58884

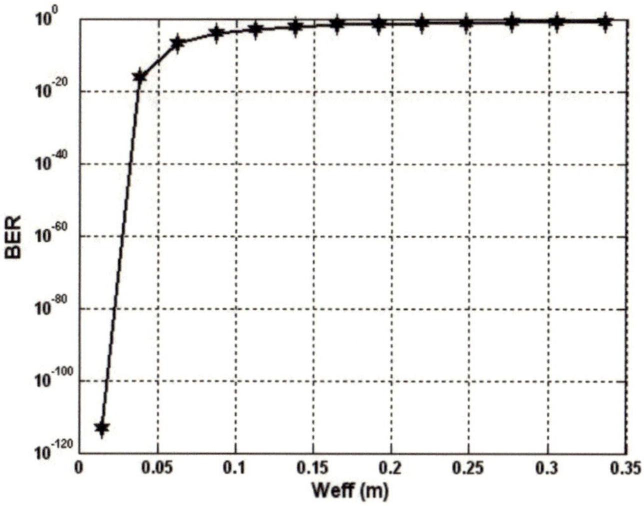

Figure 36. BER versus waist of beam with turbulence.

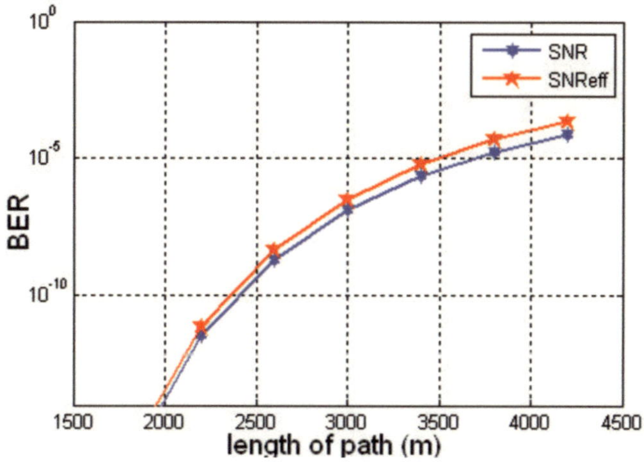

Figure 37. BER versus transmission range for SNR and $SNR_{eff.}$

7. Summary

Recently, telecommunication and computer networking are moving toward optics communication. Because the light is the hasten medium to transmit, the huge information, it's cost effective, flexibility, quick deployment, and the promise of optical bandwidth.

FSO is considered as alternative choice that can be employed as a reliable solution to broadband short distance applications. However, mobile operators are planning to use this technology to short distance links. In this chapter, we briefly introduced the concepts of FSO technology, mathematical approach of this technique. Practical part, we take the climate effects for deployment of FSO in Yemeni territorial as a case study. We have studied in detail the total attenuation influencing FSO systems. The total attenuation in this case study depends on two parameters: scattering and geometric losses. This work was concentrated on two different designs as demonstrated in Table 6. These results showed that when the link range, divergence angle, and transmitter aperture were increasing, the geometric loss increased too. But, we found that geometric loss decreased with the increasing of the receiver aperture diameter. Total attenuation also increased with increasing of the distance link, low visibility, and with decreasing of the wavelength. It was also shown that the effect of rainfall on the FSO system performance was so small that we can neglect it. In general, FSO performance was bad in Taiz at the low visibility compared to Sana'a and Aden. However, in the average visibility, the FSO performance was effective in three cities.

We concentrate on the scintillation effects on the performance of FSO links. The analysis was carried out for the variance, SNR, and BER in the environment of Yemen. Scintillation for the Yemeni environment is wavelength and distance dependent. The wavelength of 1550 nm turned out to be interesting since it is less sensitive to atmospheric turbulence and harmless to the human eye. The results indicate that the performance of the FSO system is good during the worst conditions in Yemen. To improve the transmission efficiency of FSO systems, the wavelength of 1550 nm must be used and the distance between transmitter and receiver must be reduced. To achieve a BER of 10^{-9} during air turbulence, the distance between transmitter and receiver should be 2600 m. Thus, the FSO system may be applied in Yemeni territorial efficiently even in case of the presence of air turbulence.

8. Important questions related to this chapter for self-evaluation:

1. What is free-space optical communication?
2. What is the maximum speed can FSO products offer?
3. What are the operating wavelengths of FSO systems?
4. What are the FSO technology applications?
5. Do you need a license to operate FSO system?
6. What are the advantages and disadvantages of FSO technology?
7. What is the recommended distance of the FSO link?
8. What is the different between FSO and fiber-optics?
9. Is FSO safe for eyes and human body?
10. Is FSO transmission secure?
11. Can FSO systems operate through windows glass?
12. Is the sunlight influence on FSO link?

http://dx.doi.org/10.5772/58884

13. What about the climate effect on FSO performance?
14. In figure 5, what is meant by free space and how to calculate its losses?
15. Physically, discuss the Beer`s law.
16. Calculate the antenna gain for FSO product if the wavelength is 1550 nm?
17. What is meant by divergence angle?
18. How to connect the FSO links to the network?
19. Is FSO characterized as cost effective and why?
20. Is FSO only deployed on rooftops?
21. What are the advantages of using infrared communication instead of other radio relay links line of sight (LOS)?
22. What are the transmitted power recommended for FSO products?
23. Could you design an FSO link and calculate the link budget?
24. Calculate the received power of FSO link demonstrated in figure 6?
25. What is meant by focal length?
26. If the observer is looking at a telecommunication tower of 20 meter high at a distance of 120 m and the distance between the optical center of the eye lens and fovea is of 17 mm.
    a. Graphical represent of the eye looking at a tower.
    b. calculate the size of retinal image of the tower (h) that's reflected primarily in the area of the fovea.

answer: $h=2.83\ mm$

1. What happens if a bird or an micro airplane flies through the transmitted beam?
2. Using MATLAB, Could you write the code of figure 33 and display all the curves?
3. What is meant by BER, MSE, SNR, and PSNR?
4. Is FSO technology suitable to transmit information from the earth to the satellite?

Author details

Abdulsalam Ghalib Alkholidi* and Khaleel Saeed Altowij

*Address all correspondence to: abdulsalam.alkholidi@gmail.com

Faculty of Engineering, Electrical Engineering Department, Sana'a University, Sana'a, Yemen

References

[1] fSONA unveils 2-5-Gbps free-space optical systems. September 5, 2012.

[2] Kim, Issac I. and Eric Korevaar. Availability of Free Space Optics (FSO) and Hybrid FSO/RF Systems. Optical Access, Incorporated; 2002.

[3] Rafael C. Gonzalez, and Richard E. Woods. Digital Image Processing. Prentice Hall, 3d Ed, 2009.

[4] Free space optics system design. LightPoint-White Paper Series, 2009.

[5] Heinz Willebrand, and Baksheeesh S. Ghuman. Free-Space Optics: Enabling Optical Connectivity in Today`s Networks. SAMS Publishing; 2002.

[6] Joseph Goodman. Introduction to Fourier Optics. McGraw-Hill, 2005. ISNB-0974707724, 9780974707723

[7] Roberto Ramirez-Iniguez, Sevia M. Idrus and Ziran Sun. Optical Wireless Communications IR for Wireless Connectivity. Taylor & Francis Group, Book, CRC Press, 2007. ISBN-13: 978-0-8493-7209-4

[8] Hamid Hemmati. Near-Earth Laser Communication. CRC Press; 2008. ISBN-13: 978-0-8247-5381-8.

[9] I. I. Kim, B. McArthur, and E. Korevaar. Comparison of Laser Beam Propagation at 785 nm and 1550 nm in Fog and Haze for Optical Wireless Communications. Proc. SPIE, 4214, pp. 26-37; 2000.

[10] S. G. Narasimhan and S. K. Nayar. Vision and the Atmosphere; 2007.

[11] Olivier Bouchet et al.. Free-Space Optics: Propagation and Communication. Book, ISTE; 2006.

[12] M. Gebhart, E. Leitgeb, and J. Bregenzer. Atmospheric Effects on Optical Wireless Links Presented at 7th International Conference on Telecommunications (ConTEL), pp. 395–401, Zagreb, Croatia; 2003.

[13] B. Naimullah, S. Hitam, N. Shah, M. Othman and S. Anas. Analysis of the Effect of Haze on Free Space Optical Communication in the Malaysian Environment: IEEE; 2007.

[14] Willebrand H A, Ghuman B S. Fiber optic Without Fiber. Spectrum, 38(8): 40–45, IEEE; 2001.

[15] Alkholidi A, and Altowij K S. Effect of Clear Atmospheric Turbulence on Quality of Free Space Optical Communications in Western Asia. In: Das N, ed, Optical Communications Systems, p41-74. Rijeka, Croatia: InTech; 2012.

http://dx.doi.org/10.5772/58884

[16] Arnon S.. Optical Wireless Communications. Encyclopedia of Optical Engineering, New York; 2003.

[17] Potenza Robert. Technology Update: Lighting up the Last Mile with Optics. Network World; July 22, 2002.

[18] N. J. Veck. Atmospheric Transmission and Natural Illumination (visible to microwave regions). GEC Journal of Research, 3(4), 209–223; 1985.

[19] M. A. Bramson. In Infrared Radiation. A handbook for Applications, Plenum Press, p. 602; 1969.

[20] Earl J. McCartney. Optics of the Atmosphere: Scattering by Molecules and Particles. Wiley & Sons, New York; 1997.

[21] M. S. Awan, et. al. Characterization of Fog and Snow Attenuations for Free-Space Optical Propagation. Journal, Vol. 4, No. 8; 2009.

[22] B. Bova, and S. Rudnicki. The Story of Light. Sourcebook; 2001.

[23] P. P. Smyth et. al. Optical Wireless Local Area Networks Enabling Technologies. BT Technology Journal, 11(2), 56–64; 1993.

[24] Hill, S. L. and Liu M.. Free Space Point to Point Laser and Optical Communication; 2001.

[25] Achour M.. Simulating Atmospheric Free-Space Optical Propagation part I, Haze, Fog and Low Clouds, Rainfall Attenuation. Optical Wireless Communications. Proceedings of SPIE; 2002.

[26] I. Kim, R. et. al..Wireless optical transmission of Fast Ethernet, FDDI, ATM, and ESCON protocol data using the Terra Link laser Communication System. Opt. Eng., 37, 3143-3155; 1998.

[27] Kim I I, Korevaar E. Availability of free space optics (FSO) and hybrid FSO/RF systems. Proc. SPIE 4530, Optical Wireless Communications IV, 84; November 27, 2001, 4530: 84–95. doi:10.1117/12.449800

[28] J. Li, and M. Uysal. Achievable Information Rate for Outdoor Free Space Optical. Global Telecommunications Conference, Vol.5, p2654-2658; 2003.

[29] X. Zhu, and J. M Kahn. Free-Space Optical Communication through Atmospheric Turbulence Channels, IEEE, Vol.50,No.8, p 1293-1300; 2002.

[30] Fried, D. L.. Limiting Resolution Down Through the Atmosphere. J Opt. SOC. AM Vol., 56 No 10; 1966.

[31] R. R. Beland. Propagation Through Atmospheric Optical Turbulence, in The Infrared and Elctro-Optical systems. Handbook, F.G. Smith, (Ed)., SPIE Optical Engineering press, Bellingham, WA; 1993, Vol. 2, Chapter 2.

[32] G. C. Valley. Isoplanatic Degradation of Tilt Correction and Short-term Imaging Systems. Appl. Opt. 19, 574–577; 1980.

[33] L. C. Andrews and R. L. Phillips. Laser Beam Propagation through Random Media. SPIE Optical Engineering; 1998.

[34] Xuan Tang. Polarisation Shift Keying Modulation Free-Space Optical Communication Systems. Ph.D thesis, University of Northumbria at Newcastle; February, 2012.

[35] Mazin Ali A. Ali A. Atmospheric Turbulence Effect on Free Space Optical Communication. International Journal of Emerging Technology in Computational and Applied Sciences (IJETCAS), 5(4) ; June-August, 2013, p345-351. www.iasir.net.

[36] Hennes Henniger, and Otakar Wilfer. An Introduction to Free-space Optical Communications. Radioengineering, Vol. 19, No. 2, p 203-212; June 2010.

[37] Altowij K. S., Alkholidi A, Hamam H. The effect of Clear Atmospheric Turbulence on the Quality of the Free Space Optical Communications in Yemen. Frontiers of Optoelectronics in China; 2010, 3(4): 423–428 doi:10.1007/s12200-010-0123-8

[38] Abdulsalam G. Alkholidi and Khalil S. Altowij. Climate Effects on Performance of Free Space Optical Communication Systems in Yemen. Springer, Frontier of Optoelectronics; 2014, doi: 10.1007/s12200-014-0392-8.

[39] Civil Aviation and Meteorology Authority (CAMA) data recorded report; 2008.

Chapter 6

Reliable Communication in Cooperative *Ad hoc* Networks

Unai Hernandez-Jayo,
Aboobeker Sidhik Koyamparambil Mammu and
Idoia De-la-Iglesia

Additional information is available at the end of the chapter

http://dx.doi.org/10.5772/59041

1. Introduction

Although nowadays wireless networks are a regular and familiar framework for sharing information among devices, the way in which these nets are organized and managed is evolving day by day due to the requirements of the scenarios in which they are deployed. Since those first experiments carried out by the WECA (Wireless Ethernet Compatibility) association in late 90s, the application areas and use cases in where wireless communications are applied has been changing.

Many of the wireless networks that we use daily at home, at the office or when se use a cellular are based on those first approaches, in which an Access Point is needed to have connectivity. These setups are called 'Infrastructure Mode' and use a fixed and wired backbone to address information from the source AP to the destination AP. But in some situations these networks are limited by their own nature due to their need for an AP, a base station, some routers or switches and so on. It is in these scenarios where a ''Infrastructure-less Mode'' can overcome these drawbacks, allowing the nodes of a network to routing and forwarding information for other nodes, without relying on centralized administrator. These types of networks are called wireless ad hoc networks [1].

Now, if we have into consideration the current trends in technology, it can be said that mobility and ubiquity are common characteristics to all the new gadgets launched to the market. Users want to be online anytime and everywhere and to obtain information from all the surrounding elements. Then, we talk about Mobile Ad hoc Networks (MANETs), that is, wireless networks with a dynamic shape, a shifting number of nodes, a defined bandwidth and other character-

Figure 1. Infrastructure based networks (left) and Ad hoc networks (right)

istics, where the nodes can be any kind of devices with communications and networking capability that communicate with each other without a centralized coordinator [2]. In this scenario, each node can play the role of a router, hosting the network topology dynamically, because as it was mentioned above, the shape and the topology of the net can change as well as the nodes on it. The main characteristics of MANETs can be summarized as follows [3]:

- Dynamic topologies: network topology can change quickly due to the nodes can move freely in the net.
- Bandwidth constrains: compare with wired networks, the capacity of a MANET is relatively small and also it is sensitive to interferences, noise, and signal fading effect.
- Energy constrains: although many of the nodes can be plugged to the power line or they can be equipped with big batteries, some of them use small power supplies, so during the network design it is necessary to consider how to save power in order to assure the stability and longevity of the network.
- Limited physical security: although the decentralized nature of MANETs provides robustness against the single points of failure, these nets must be protected against eavesdropping, spoofing, and the injection of malicious data attacks.

In this context, thanks to the rapid increase and improvement of the mobile computing a wide set of wireless devices have proliferated, making possible that traditional hardware as digital cameras, thermostats, cooking ovens or washing machines are provided with communications and computing functionalities so they can be part of a MANET. This new paradigm is known as Internet of Things [4], that is, a scenario in which all the objects beyond computers, mobiles or touch screens have the ability of generating, sharing and processing information in a pervasive manner [5]. With all of this, technologies must have evolved to new standards, architectures, protocols, hardware, services and facilities that will make possible a control of the way in with all the nodes access to the net to share their information.

One scenario that represents perfectly the characteristics and it is a perfect case of study of MANETs is the Vehicular Ad hoc Networks (VANETs), a subset of MANETs, which creates wireless networks between vehicles [6]. In a VANET each vehicle is a moving node which creates wireless networks with surrounding vehicles [7], thanks to the On-Board Unit (OBU), a hardware with communications and computing capabilities that allows drivers to receive information about events that can affect his driving. Then, the main function of the OBU is to exchange information with other vehicles or Road Side Units (RSUs), elements located at the

http://dx.doi.org/10.5772/59041

infrastructure that act as gateways between the VANET and other networks or agents as Traffic Management Centers (TMC). These centers are placed far away the VANET and play an important role in the applications developed in the area of VANETs, coordinating the information that is shared among VANETs that are deployed in different geographical areas.

In VANETs they can be distinguish two types of links: vehicular-to-vehicular communication (V2V), based on an Ad hoc architecture, vehicles exchange directly messages without a central coordinator; and vehicle-to-infrastructure or infrastructure-to-vehicle (V2I or I2V), where the messages are shared between the vehicles and the RSUs. VANETs are designed for a huge range of cooperative applications, that is, services that provide information to the drivers thanks to the data shared among all the vehicles on the net. These can be safety and non-safety applications, which allow several added services as infotainment, traffic management, toll payment, and geographical based services and so on [8]. That is, VANETs make possible to deploy applications that help to improving the transport services and traffic conditions using collaborative systems based on V2X Ad hoc networks.

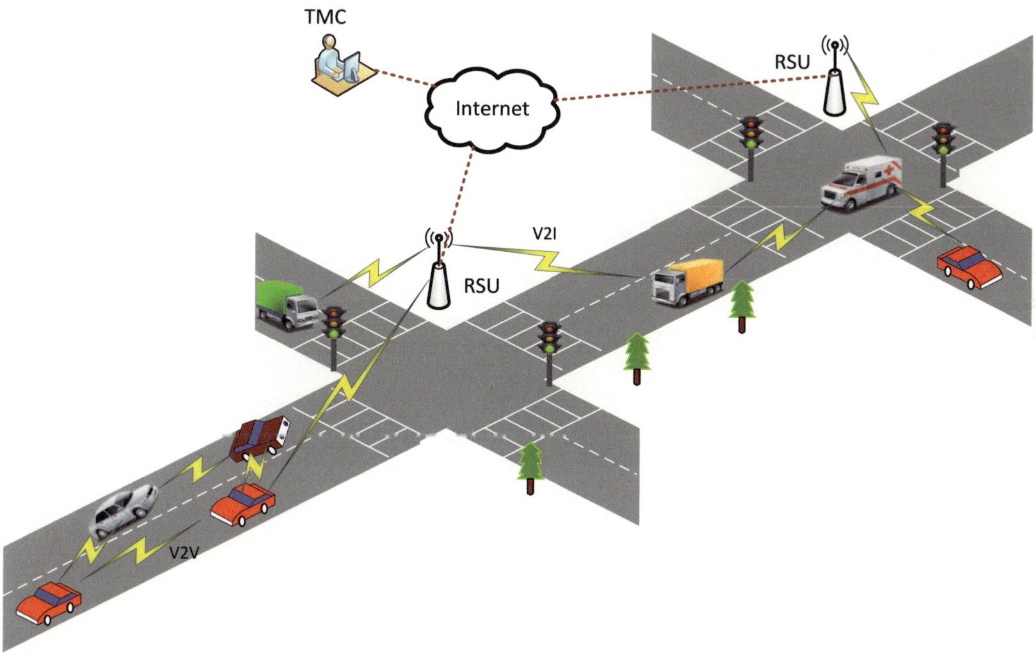

Figure 2. Vehicular Ad hoc Network Scenario

This introduces the definition of Intelligent Transport Systems, where each vehicle is a sender, a receiver and a router at the same time, so it can broadcast the information to the VANET, which uses this information to provide these safety and non-safety services to the drivers. The OBU is the hardware in charge of processing these data and it also enables these short range wireless ad hoc networks (the coverage area is around 300 meters) but it also must dispose other systems that permit to report position information such as Global Positioning System (GPS) or a Differential Global Positioning System (DGPS) receiver if more accuracy position information is required. This information is quite important because most of the services that are available in a VANET depend on the geographical position of the source and the destination. Table 1 presents a classification of ITS applications that can be deployed using the VANET architecture [9].

Category	Applications	Uses cases
Active safety applications	Cooperative driving assistance applications	Emergency vehicle warning Slow vehicle indication Intersection collision warning Motorcycle approaching indication Emergency electronic brake lights Wrong way driving warning Stationary vehicle - accident Stationary vehicle - vehicle problem Traffic condition warning Signal violation warning Roadwork warning Collision risk warning
Efficiency applications	Traffic management Road monitoring	Regulatory / contextual speed limits notification Traffic light optimal speed advisory Enhance routing Road conditions sensing (rain, visibility, wind, hazardous location, road adhesion)
Infotainment applications	Contextual information Entertainment	Point of Interest notification Automatic access control and parking management ITS local electronic commerce Media downloading Insurance and financial services Fleet management Loading zone management

Table 1. ITS applications on VANETs

These applications can be deployed on urban or motorway scenarios, each one with its own particularities. In an urban scenario, many of the times there is not line of sight between the nodes so fading and communication disruptions are frequents. In a motorway, the high density of vehicles can overload the radio channels in which the VANETs work. Yes, although maybe users are not aware about that, the radio spectrum (the physical interface used by wireless communications networks) is a limited resource that it must be shared among all OBUs and RSUs that shape the VANET. Commonly, ISM (Industrial Scientific Medical) radio bands with

http://dx.doi.org/10.5772/59041

frequency ranges 2.40–2.4835 GHz and 5.15–5.875 GHz are used by wireless networks for license-free communications [10]. The definition of these standards is crucial in order to attend the increase on the demand of the spectrum channels and to make possible that different networks can coexist in the same radio band.

Although WLAN (IEEE 802.11a/b/g/n) could be the technology used in VANETs, most of the applications included at Table 1 require time-critical communications, a continuous handover among different RSU in V2I/I2V links, and as these standards use CSMA (Carrier Sense Multiple Access), so many of the nodes cannot have success in channel access due to the high density of some scenarios. Due to the limitations of these standards in mobile scenarios as VANETs, a new extension has been developed: IEEE 802.11p, designed specifically for vehicular environment in which high reliability and low delay characteristics are mandatory. This new standard, known as Dedicated Short Range Communication (DSRC) uses the physical layer of IEEE 802.11a working on the 5.9 GHz band and quality of service enhancements of IEEE 802.11e. Network and transport layers are in the scope of WAVE (Wireless Access in the Vehicular Environment) standard which defines the protocols and services that support multi-channel wireless connectivity between IEEE 802.11 Wireless Access in Vehicular Environments devices [11].

Once the access to the medium is defined under the frame of the IEEE 802.11p standard, in a situation in which many nodes have information to transmit to different destinations in a network that is geographically distributed, it is quite important to determine the protocols that allow to organize the addressing of the information and to assure that all the nodes have the chance of transmitting and receiving data. The nature of MANETs, and specifically of VANETs implies that the maintenance, management and routing task of the network must be done by all the nodes, making these kind of networks more difficult or more complex to other wireless networks. Therefore, advances techniques of management and arrangement should be applied to organize the network and assure its effective implementation and its fairness and reliability for all the nodes.

In the next sections of these chapter are analyzed the main techniques used to disseminate data in VANETs, with an special emphasis in clustering, a control scheme that can take into consideration the speed and distance difference among neighboring nodes in the VANET to group them in order to assure a stable cluster structure and then enhance the stability of the network topology.

2. Data dissemination algorithms

Data dissemination in VANETs has recently received considerable attention. Due to the unique characteristics of VANET, the implementation of reliable data dissemination among vehicles has encountered many challenges. Information dissemination in VANETs provides drivers a way to be aware in real-time of everything that is happening in their surroundings. A wide range of information can be disseminated, including traffic and road conditions, closure and detour information, incident information, emergency alerts, and driver advisories.

Information dissemination schemes in VANETs are commonly categorized into two different groups, according to the type of ITS application: safety and non-safety. During the last years, research community has focused their studies more on safety applications which are highly demanding in terms of message delay and present a challenging field of study. Although in safety applications the frequency of messages is low, the message delay is a key factor because a safety message, e.g., an emergency vehicle warning, has to reach a maximum number of nodes in a given area within a very short time interval, because after this time interval, the message essentially becomes useless.

However, in non-safety applications the message delay loses importance since the message could be useful for a longer time interval, even up to several minutes, e.g., for disseminating traffic road conditions. On the other hand, the frequency of these messages is much higher in this type of applications.

Therefore, data dissemination in VANET is a challenge for the deployment of cooperative services and applications because the dissemination routing protocol has to be suitable both for safety and non-safety applications, and it also has to be aware of the vehicular environment challenges as the high mobility of nodes and the extremely dynamic network topology. Therefore, the design of an efficient information dissemination routing protocol for VANETs is very crucial.

The function of a routing protocol in Ad-Hoc network is to establish routes between different nodes and the main requirement is to achieve minimal communication time with minimum consumption of network resources. The main reasons that make so difficult the design of these routing protocols are the highly dynamic nature of VANETs due to the high mobility of the nodes, and the need to operate efficiently with limited resources, such as network bandwidth. Moreover, routing protocols in VANETs, and generally in every Ad-hoc Networks, are not so good in scalability due to frequently changing network topology, lack of predefined infrastructure and limited radio communication range. In the literature, four categories of dissemination routing protocols for VANETs which are presented: position-based, broadcast, geocast and cluster-based.

Broadcast routing is commonly used in ITS applications in VANETS because it guarantees that every vehicle will receive the message. The simplest way to implement a broadcast service is flooding in which each node re-broadcasts messages to all of its neighbors except the one it got this message from. Flooding performs relatively well for a limited small number of vehicles and is easy to be implemented. Furthermore, this protocol is very reliable in safety applications but it consumes high bandwidth and resources, and it can also provoke a broadcast storm when the number of nodes in the network increases. If multi-hop communications are implemented as each node receives and broadcasts the message almost at the same time, this routing protocol generates contentions and collisions and high bandwidth consumption.

However, there are many studies where they use broadcast, but they design an approach to avoid broadcast storm. In [12], Yang et. al propose a V2V communication protocol for Cooperative Collision Warning application. In this approach when a vehicle has an incident, it becomes an abnormal vehicle (AV) and starts broadcasting periodically Emergency Warning

http://dx.doi.org/10.5772/59041

Messages (EWMs), with its geographical position, speed and direction to its surrounding vehicles. If this incident provokes that more vehicles have to stop and, therefore, they become also AV, only one of them is going to send the EWMs to avoid the broadcast storm. In [13], Ferrari et. al use broadcasting protocol with multi-hop communication but to avoid the broadcast storm not every vehicle forward the received messages, only the farthest vehicles from the source forward it.

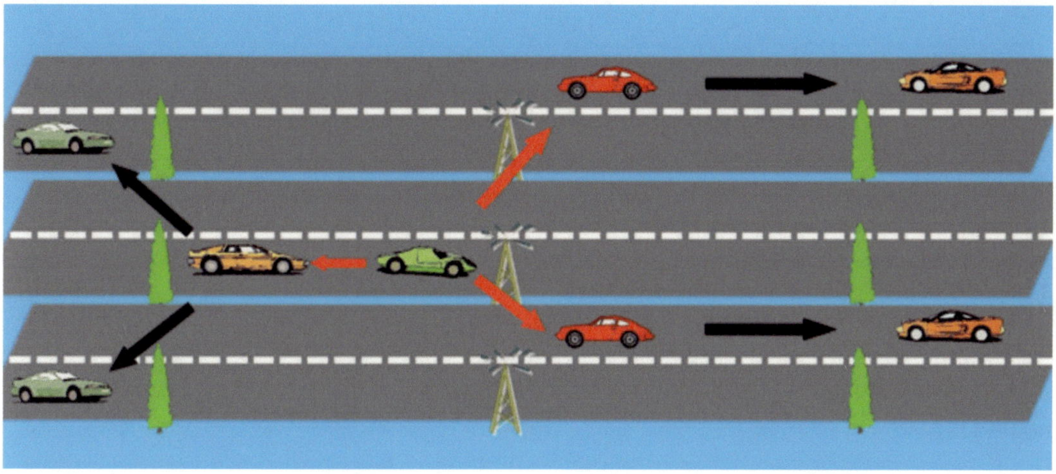

Figure 3. Broadcasting routing protocol

In the position-based routing protocol the forwarding dissemination decisions are based on location information. This approach makes sense because in VANETs the movements of the vehicles are usually restricted in just bidirectional movements constrained along roads and streets, and the geographical location information of vehicles is taken from street maps, traffic models or even more prevalent navigational systems on-board the vehicles. This protocol is commonly used with multi-hop communications and therefore, nodes usually forward the packet to a node that is geographically closest to the destination. The main advantages of this routing protocol are:

- It does not require routing tables
- Traffic overhead may be small
- Supports delivery of packets to a geographical area, called geocasting

For example, as it is shown in Figure 4, if one vehicle has an accident the information will be only be necessary for the vehicles that are behind the damaged vehicle, not for the ones that are not going to drive again though the point the accident has happened.

However, to use this location-based routing protocol in a built-up city environment is very challenging, due to vehicles are distributed in an irregularly way because they usually are more concentrated on some principal roads than others and the road patterns define their

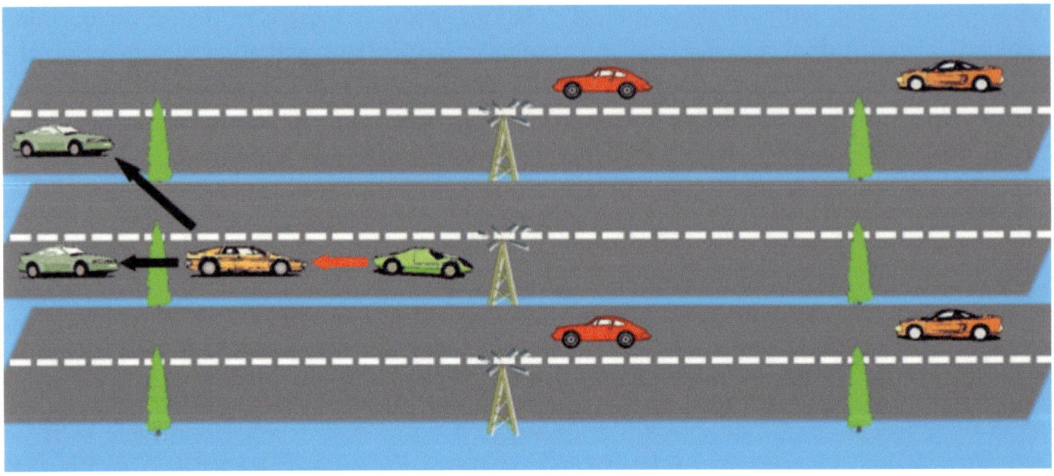

Figure 4. Position-based routing protocol

mobility and add difficulty in the signal reception because of the radio obstacles such as high-rise buildings which may lead VANETs unconnected. Furthermore, in general, topology-based routing protocols are considered not to scale in networks with more than several hundred nodes [14].

In order to position-based routing protocol could work, vehicles should send periodically beacon messages to announce their position and enable other nodes to maintain a one-hop neighbor table. This approach is scalable and resilient to topology changes since it does not need routing discovery and maintenance; however, periodic beaconing creates a lot of congestion in the network [15]. This beaconing frequency can be configured according to different scenarios or traffic situations, but if this beaconing frequency is not enough the inaccuracy of position information is higher and a neighbor selected as a next hop may no longer be in transmission range implying to a significant decrease in the packet delivery rate.

Therefore, the key ideas we have to take into account to select one position-based routing protocol are:

- Loop-freedom: routing protocols should be inherently loop-free and should avoid recovery strategies using timeouts of old packets and memorizing packets that have been seen before
- Distributed operation
- Path strategy
- Metrics
- Memorization
- Guaranteed delivery
- Scalability
- Robustness

http://dx.doi.org/10.5772/59041

There are three different kinds of position-based protocols which are restricted directional flooding, greedy and hierarchical routing protocols. The most used routing position-based protocol is the greedy in which they use forwarding to route packets from a source to the destination. This strategy do not establish and maintain the routes between the source and the destination; on the other hand, a source node define the approximate position of the destination and add this data in the data packet and selects the next hop depending on the optimization criteria of the algorithm; for example, as it is shown in Figure 5, one criteria could be the closest neighbor to the destination [16],[17]. In the same way, each intermediate node selects a next hop node until the packet reaches the destination, as it is shown in Figure 4 Position-based routing protocol.

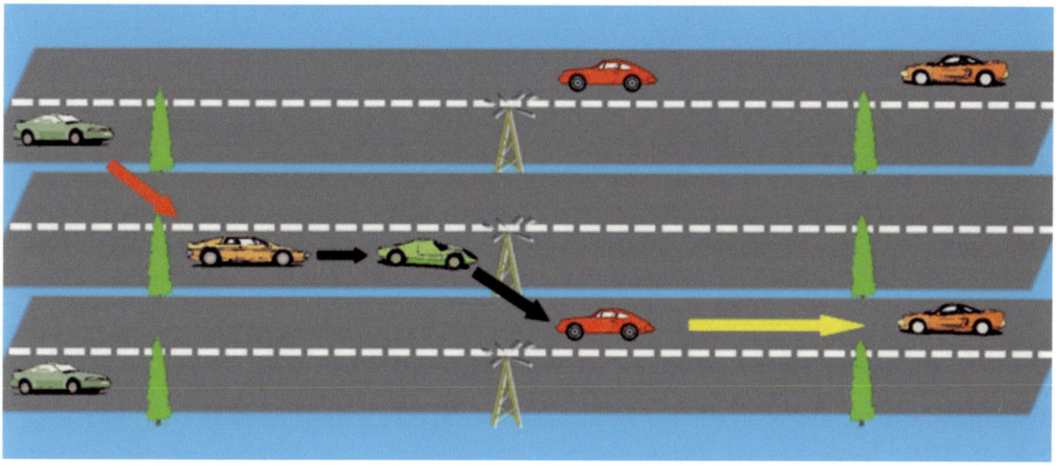

Figure 5. Greedy routing protocol

The main characteristics of Greedy algorithms are:

- Loop free
- Localized information
- Single path strategy
- Metric: Hop count
- No memory
- No guarantee of delivery
- Scalable
- Somewhat robust

In restricted directional flooding, the sender will broadcast the packet to all single hop neighbors towards the destination. The node which receives the packet checks whether it is within the set of nodes that should forward the packet (according to the used criteria). If it is, it will forward the packet. Otherwise the packet will be dropped. In restricted directional flooding, instead of selecting a single node as the next hop, several nodes participate in forwarding the packet in order to increase the probability of finding the shortest path and to be robust against the failure of individual nodes and position inaccuracy.

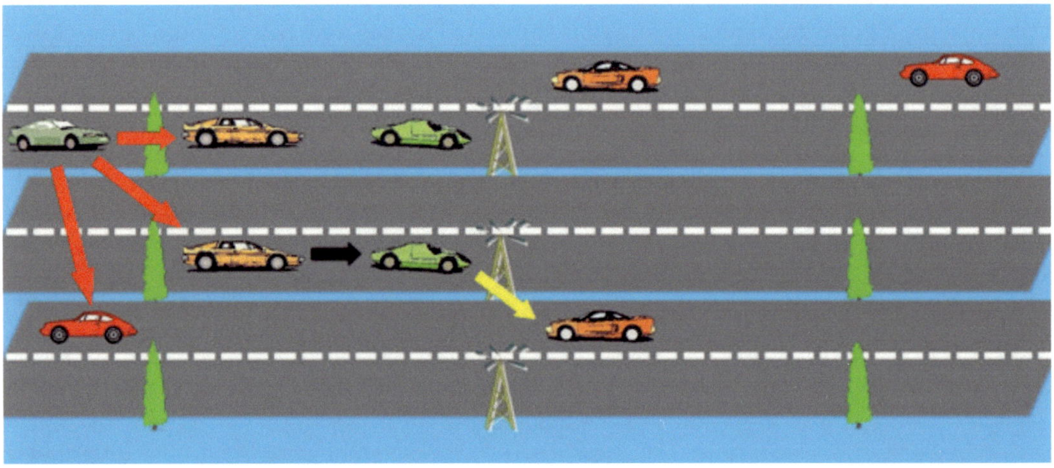

Figure 6. Restricted directional flooding routing protocol

The main characteristics of Restricted Directional Flooding are:

- Not loop free
- Localized operation
- Path strategy: flooding / multipath
- Metric: Hop count
- Memory
- No guarantee of delivery
- Not scalable
- Not robust

The third forwarding strategy is to form a hierarchy in order to scale to a large number of mobile nodes. This strategy tries to reduce the complexity of the information each vehicle has to handle and also improves the scalability of the network. The two main strategies used to combine nodes location and hierarchical network structures are the zone-based routing and the dominating set routing [18].

Geocast routing is a location-based routing but in a multicast way, so each message is broadcasted to every vehicle inside a defined area. In Figure 7 it is shown that the defined area are the vehicles which receive the yellow messages. Geocast can be implemented with a multicast service by simply defining the multicast group to be the certain geographic region.

http://dx.doi.org/10.5772/59041

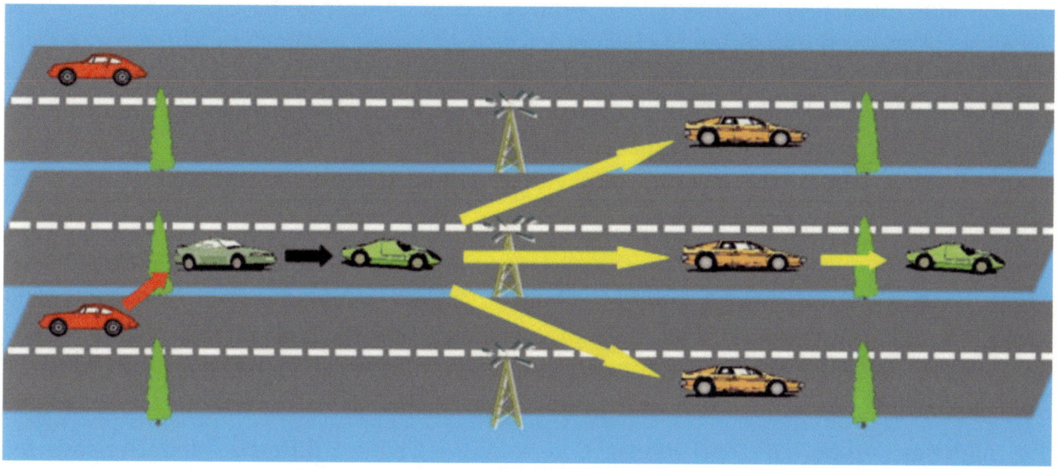

Figure 7. Geocast routing protocol

Most geocast routing methods are based on directed flooding, which tries to limit the message overhead and network congestion of simple flooding by defining a forwarding zone and restricting the flooding inside it. With this routing protocol we consume less network resources than broadcast routing but we also guarantee that every vehicle will receive the message. However, we continue having the broadcast storm problem unless we only use one-hop communications. Geocast routing is divided into three types which are: Routing with simple flooding, direct flooding and no flooding [19].

The Geocast routing based on simple flooding was not created for geocast routing but it is used as a basic unit and for the comparison with other protocols. In this method, the source vehicle delivers the packet to all other nodes in the network and all receivers have to check whether they are within the destination area. This is a very straightforward approach but is not a well-organized approach. In this approach, information of location is not used.

In the Geocast routing based on direct flooding the packet is forwarded to a defining region called "forwarding zone". In this approach a packet is only forwarded to forwarding zone by the source node and not to all nodes in the network. In other words, this protocol is based on flooding but avoids flooding the whole network by defining a forwarding zone, and therefore, outside the forwarding zone the packet is discarded. There are two types of forwarding zone, the first one is the rectangular forwarding zone and the other one is distance-based forwarding zone.

The Geocast routing without Flooding is a simple geocast routing protocol that uses a regular unicast routing protocol between the sender and the destination region. Inside the destination region, flooding can be used, as well as any other routing protocol that can be independent of the protocol used outside the destination region, but the main difference is that it does not use flooding outside the forwarding zone.

But the most used routing protocol for vehicular environment is the cluster-based, where vehicles are grouped into different clusters according to some parameters. These parameters differ from one algorithm to another and are the key factor to build stable clusters. Some of those parameters could be the location, speed or inter-vehicle distance. Other parameters, as the IEEE 802.11p wireless coverage area of each vehicle, could affect in the size of clusters which could vary from one cluster to another in the same network depending on the location of nodes.

Therefore, clusters are virtual groups selected by a clustering algorithm where at least there is Cluster Head (CH) and some Cluster Members (CMs). The main advantage of cluster-based solution is that it can achieve good scalability for large networks, but, on the other hand, the delay and overhead involved in the formation and maintenance of clusters has to be taken into consideration.

The highway, urban, city and intersection scenarios require different characteristics for selection of CHs and for formation of clusters.

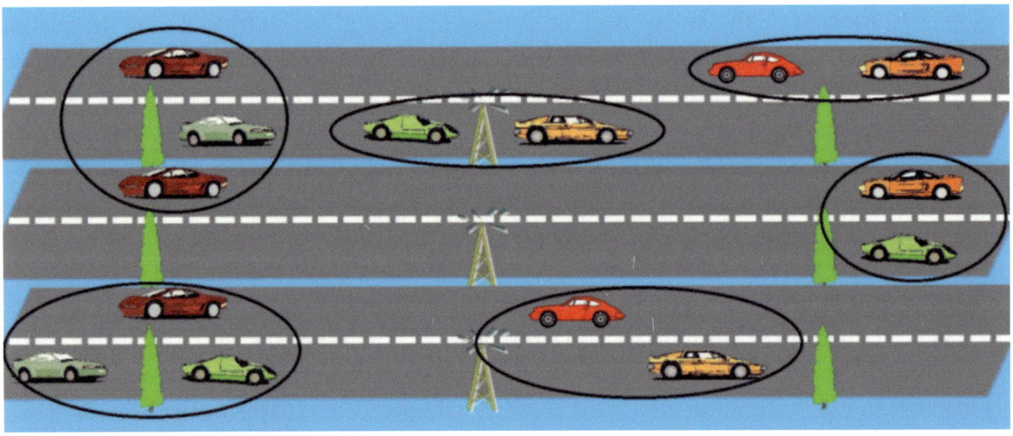

Figure 8. Clustering routing protocol

The cluster-based routing solution could be designed in three different ways depending on how vehicles discover the CH. It could be in a proactive, reactive or hybrid way. In the proactive solution beacon messages are constantly broadcast and flooded among vehicles since every vehicle should maintain updated their neighbor table to know which the next hop node toward a certain destination is. The advantage of the proactive routing protocols is that there is no route discovery since route to the destination is maintained in the background and is always available upon lookup. Despite its good property of providing low latency for real-time applications, the periodically beacon sending for the maintenance of the neighbor table requires a significant part of the available bandwidth, especially in highly mobile VANETs.

In the reactive approach the configuration phase is initiated by the vehicle because it starts a communication when it needs to communicate with another vehicle. It maintains only the routes that are currently in use, thereby reducing the burden on the network. Reactive routings

http://dx.doi.org/10.5772/59041

typically have a route discovery phase where query packets are flooded into the network in search of a path. The phase completes when a route is found.

In a mixed approach vehicles also send periodic proactive beacon messages to have the neighbor table updated but they are also able to create a new communications on demand when they need to send any message to another vehicle.

To sum up, it is not very obvious which is best routing protocol for data dissemination in VANETs because it depends on application and the characteristics of the scenario like the position of the vehicles, speed, direction of movement, potential communication duration and potential number of communication neighbours, among others. Therefore, research community should continue researching on the development of new dissemination data routing protocols.

3. Clustering algorithms

Clustering is a technique for grouping vehicles in the geographical vicinity together, making the network more robust and scalable. Under a cluster structure from Figure 9, vehicles may be assigned a different status or function, such as cluster head (CH), gateway (GV), or cluster member (CM). A CH normally serves as a local coordinator for its cluster, performing intra-cluster transmission arrangement, data forwarding, and so on. A GV is a non-CH vehicle with inter-cluster links, so it can access neighboring clusters and forward information between clusters an RSUs. A CM is usually called an ordinary vehicle, which is a non-CH vehicle without any inter-cluster links.

Cluster-based solutions may be a realistic approach in supporting reliable and scalable multi-hop communication for VANETs [20]. Clustering has been shown to effectively reduce data congestion [26], and can support Quality of service (QoS) requirements [21] for both delay-tolerant (e.g. road/weather information) and delay-intolerant (e.g. safety messages). According to [22] clustering provides three basic benefits.

- Spatial reuse of network resources.
- Emergence of a virtual backbone.
- Improved network stability and scalability from the viewpoint of regular CMs.

Clustering can be done in a centralized or decentralized way. In centralized way, RSU elect CHs and forms clusters based on periodic message. As a fixed infrastructure, the RSU should be fully utilized to collect information and use this information to perform central control. It acts as backbone of all data transmissions. However, it does not work in network where there are no RSUs. Decentralized clustering is based on the "hello message" exchange between the vehicle and it forms clusters and elects its CHs. Additionally, most protocols only use peer to peer communication to gather and transmit information, so those data can hardly be converged and processed in centralization. This is further discussed in detail in coming section.

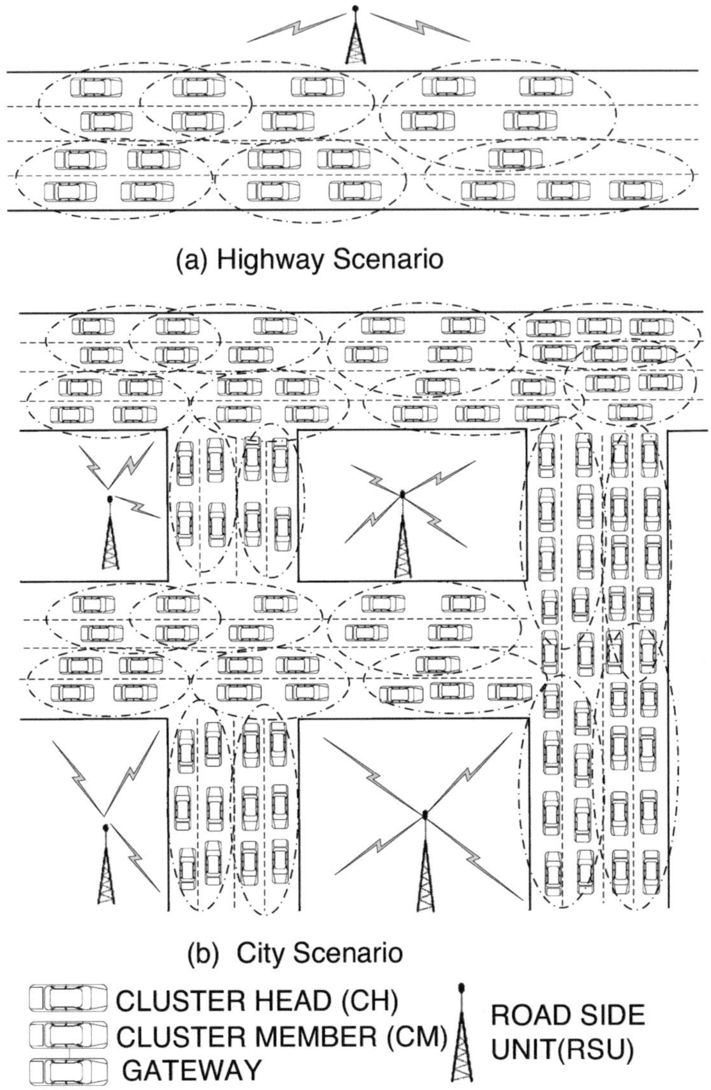

Figure 9. (a) Highway and (b) City scenario.

3.1. Infrastructure centric clustering

Infrastructure based clustering is a centralized clustering where it gathers information from all the vehicles in the road, including speed, direction, positions, and further traffic related information. Infrastructure divides vehicles in the road into different cluster groups, it coordinates in the election of CH, routing of packets and allocation of the channel to its CMs. As a fixed infrastructure, it computes the collected information to perform central control. Moreover, using V2V clustering some algorithms require additional devices for computation to fulfill the aim, which will raise the vehicles cost and reduce the feasibility of algorithms.

Infrastructure based clustering is used to solve the above-mentioned shortcomings and to achieve high stability. Overall, the amount of data to be sent is comparatively small (the position, speed, direction of each vehicle), but the communication reliability is vital.

Some approaches shown in Table 2 uses infrastructure for centralized channel allocation in order to reduce channel allotment time and control overhead. It can be seen infrastructure divides the spectrum allocated to a particular area into prefixed overlapping spatial clusters. The medium in each cluster is divided into time slots and each time slot is allocated to a vehicle in accordance to the priority of the message and availability of the time slot. However, due to centralized allotment the reliability and fairness is lowered. In another approach, infrastructure allocates channels to the moving vehicles based on their clusters and enables channel reuse in non-adjacent clusters. The infrastructure broadcast is heard by all the neighboring vehicles in the infrastructure region and this solves the issue of hidden/exposed vehicles. Furthermore, broadcast helps to avoid contention and results in efficient utilization of the allocated bandwidth. The lack of contention for channel acquisition and priority list at the infrastructure allows the protocol to ensure predictable delivery of safety messages. Nevertheless, these types of algorithms may not scale at high density and would not function in ad hoc mode in regions where there are no infrastructures.

Protocols	CH election	Cluster Formation	MAC	Scenario	Simulator	Pros	Cons
CMAC [23]	RSU is CH	Speed, relative distance and direction.	FDMA	Highway	MOVE, SUMO and NS-2	Predictable and reliable.	Density not considered. Require RSUs. Low bandwidth utilization in sparse traffic
Ranjeet Singh [24]	RSU is CH for all clusters	Static cluster formation	TDMA	Intersections	NCTUns	End to End delay is reduced.	Require RSUs all time. Reliability lowered in high speed scenarios.

Table 2. Comparison between various infrastructures based protocols.

Vehicular motion are confined to strait jacket roads and travels at high velocity and the enter/ exit infrastructure area in short interim's of time. At a given period of time, the total number of vehicles in an infrastructure area can vary significantly from a small density of vehicles to a large density of vehicles in a very short interim of time. Algorithms must be distributed or should require partial infrastructure assistance with an efficient hand-off from one infrastructure to another to meet these attributes. The vehicular movement is predetermined to road structure and directional antenna would be suitable for communication via infrastructures. The vehicle broadcast radio frequencies with transmission channels, each one considered as a common medium over which two neighboring vehicles cannot transmit simultaneously because a transmission collision occurs. So, in order to efficiently share the medium, MAC protocol is needed and is beset by contention delay. However, a protocol must ensure that safety messages are delivered within a prescribed time frame. The protocol must not suffer without the hidden/exposed terminal or deafness problem to ensure reliable message delivery. Although the infrastructure is an extra, it will be furnished on the highways extensively and applied in VANET in the near future. Therefore, compared with great and lifelong benefit, the infrastructures expense is of trifling importance at all. The efficient cluster based MAC and routing protocols can provide a more stable communication than a solution using V2V clustering. The optimum protocol should that take the advantages of fixed infrastructure and optimize the problem.

3.2. V2V based clustering

V2V based clustering is a decentralized clustering where clusters are formed based on communication between vehicles. Additionally, the CH election will be based on V2V communication. There are several advantages of using V2V-centric clustering as compared with the infrastructure-centric VANETs. V2V-centric clustering can avoid the short communication link period, high frequent hand-offs, fast channel fading, etc., that are caused by the high relative-speed difference between the fixed infrastructure and the fast-moving vehicles. Finally, the V2V-centric clustering performs better in active safety applications, which only requires exchanging messages among one hop vehicles within their transmission range.

V2V communications are expected to significantly improve transportation safety and mobility on the road. Several applications of V2V communications have been identified, from safety and warning applications, up to traffic control and driver assistance applications. In infrastructure centric clustering, all the communications is done via the infrastructure which causes a lot of control overhead and additional delay. Furthermore, it would be very cost intensive to build an infrastructure based communication all along the road structure. The V2V based clustering technique avoids the use of stationary base stations by building up VANETs, where all vehicles in a common transmission radius can exchange messages. However, CH selection carried out through V2V communications has some shortcomings, e.g., highly complex protocols, large computation and communication cost, need of additional devices and so on. Another important issue is that the connection between two adjacent CHs may be lost due to vehicles high speed, which drastically reduces the link quality. The hidden terminal problem where two vehicles are outside of each other's transmission radius, but both attempt to transmit to a vehicle that is within the radius of both. This issue is likely in pure V2V scenarios where there is no centralized communication system. The result of the hidden terminal problem is data collisions. By enabling vehicles to transmit/receive messages with each other via V2V as well as with infrastructure communications, VANETs could contribute to more safer and congestion free roads by providing correct and timely message to neighboring vehicles and other related departments.

3.3. Clustering in layers

Clustering can simplify essential functions like bandwidth utilization, routing, and channel access. In MAC layer, it can provide a fairer and reliable channel access to all vehicles in network. This can lead to increase in the reliability of packets and scalability of the network. In network layer, clustering for routing can find the closest vehicles to intended destination.

Furthermore, it reduces the number of broadcast and flooding messages in the network. In addition, the overhead for clustering is reduced if the same scheme is used for MAC and routing.

3.3.1. Clustering in MAC layer

Introducing a cluster scheme already on the MAC layer additionally provides the possibility of a fairer medium access. When clustering applied in VANETs, it brings interesting research point such as broadcast storm that occurs when several vehicles are passing at a specific region at the same time, causing network congestion, packet collisions and delays in the medium access layer. A cluster-based MAC scheme is needed in V2V communication to overcome the lack of specialized hardware for infrastructure and the mobility to support network stability and channel utilization. In this case the CH can take over the responsibility to assign bandwidth to the CMs and therefore even QoS support can be improved. As the bandwidth can be assigned centrally fewer collisions have to be expected which consequently increases the reliability.

Many researchers have proposed cluster based multi-channel medium access control protocols to improve the performance and reliability of VANETs. In these protocols, clustering is used to limit channel contention and provide fair channel access within the cluster. On the other hand, multi-channel is used to increase the network capacity by the spatial reuse of the network resources and reduce the effect of the hidden terminal problem. Moreover, to optimize the communication range and the cluster size is very difficult especially in a highly dynamic environment such as VANETs. However, in order to overcome this situation some approaches divide the service area into a set of region units, and limit the number of vehicles in each region unit for the contentions of radio channels. Each region unit is then associated with a non-overlapping radio channel pool. Since the number of vehicles in each region unit is limited, the contention period is reduced and the throughput is increased. However, these types of approach have low bandwidth utilization in case of sparse traffic. Some of clustering algorithms try to minimize the total number of clusters by creating hierarchical clusters with a diameter of at most four hops. In this section of the chapter, we compare well known cluster based MAC protocols in Table 3.

The MAC layer is divided into different cycles and each cycle is divided into contention based or contention free. In the current literature, several MAC protocols have been proposed to VANETs. Some of the well-known MAC protocols are ADHOC MAC [32], SDMA [33], VeMAC [34], DMMAC [35], STDMA [36], VeSOMAC [37] etc. These protocols are proposed for various scenarios and have many drawbacks such as hidden terminal problem, time unbounded, unreliability etc. There is a need for new MAC protocol in VANETs that can provide mobility (i.e., the MAC protocol should support vehicles to leave and join inter-vehicle communications at high speed), delay bounded (i.e., the communication must be delay bounded and real-time), scalability (i.e., VANET should scale itself according to the number of vehicles present), bandwidth efficiency (i.e., the radio resource should be utilized in an efficient and fair manner), cost (i.e., for cost-efficient and reliable communications, VANET should be fully decentralized), and fairness (i.e., every vehicle should get a fair chance to get the radio channel).The

Protocol	CH Election	Cluster Formation	MAC	Scenario	Simulator	Pros	Cons
HCA [25]	Maximum number of messages received from cluster relays is the CH.	Maximal distance between a CH and CM is two hops.	TDMA	City	OMNeT++, SUMO	End to End delay reduced.	Overhead and packet loss is increased. Do not consider the direction of movement.
Zaydoun [26]	Vehicle nearer to middle of the cluster is the CH.	Not Specified	TDMA	City	C++ with graphical interface.	Support both safety and non-safety applications.	Not suitable for high traffic. High overhead.
Xi Zhang [27]	No reception of a message longer than a particular time units from a CH, then it elects itself as CH.	RSS > threshold.	TDMA in CMs-CHs, CSMA/CA CHs-CHs	Highway	Simone 2000	Reduces data-congestion and supports safety messages.	High overhead and complex algorithm. Require two transceivers
CBMAC [28]	CH is based on waiting period of Hello messages to neighbors.	Undecided state to CM based on reception of one CH messages.	TDMA	City	I-V Communication Based on Traffic Modeling.	Minimizes the hidden terminal problem.	Does not select a stable CH during initial CH election.
RCM [29]	No CH	Geographical area. Vehicles are assigned to different channel pools.	TDMA	Highway	A. law el al	Reduced contention and throughput is increased.	Low bandwidth utilization in sparse scenarios.
TCMAC [30]	Lane weight, average distance, maximum number of neighbors, and average distance level.	Not specified	TDMA	Highway	Ns-3	Channel utilization, scalability, avoids hidden terminal problem, decreases collisions and packet drops.	Cannot be used for safety applications, it is delay intolerant.
CFIVC [31]	Random after relaying one packet to ordinary node.	According to speed.	CDMA, MCSCDMA	Not simulated	Not simulated	Avoids data collisions.	It neglects any condition that might affect the maximum speed achievable by the vehicle nodes.

Table 3. Comparison between various cluster based MAC protocols.

challenge of successfully deploying VANET services is to ensure timely and reliable data delivery for mobile vehicles.

3.3.2. Clustering in network layer

In network layer clustering, a virtual network infrastructure must be created through the clustering of vehicles sharing similar characteristics in order to provide scalability. Routing protocols for VANETs mostly based on periodical broadcast messages to reveal their positions and traffic information to neighbors. Nevertheless, deterioration of routing performance is anticipated in urban areas due to high density of vehicles in the network. Basically, excessive broadcast messages as well as broadcast overhead may increase, resulting packet losses (due to collision) and significant routing performance deterioration. Information transfer or dissemination needs multi hop communications. When exchanging information between vehicles, there are network issues that must be addressed, including the hidden terminal problem, high density, high node mobility, and data rate limitations.

In multi-hop data forwarding method, the key problem is selecting the relay/CH for data routing. Most of the relay/CH selection method presented is more suitable for highway scenarios. In a city environment, the widely adopted method is the store-carry-forward scheme. Reactive protocols find routes on-demand. If a node wants to communicate with a node to which it has no route, the routing protocol will try to establish the shortest route between them. Here there is significant delay in determining the route. Proactive (table-driven) protocol, which is based on the exchange of control packets and it is continuously updating the reachability information in the routing table, so routes are immediately available when requested but there is high overhead in maintaining updated periodic routing tables and also maintains the routes that are not going to be used. Hybrid protocol is combination of proactive and reactive protocol. It is also known as cluster based routing. It is a convenient way for developing efficient routing scheme in VANETs. In Table 4 we compare between various cluster based routing protocols proposed in VANETs.

Protocols	Cluster or CH	Scenario	Simulator	Pros	Cons
TMRC [38]	Direction of vehicle after crossing the intersection	Intersection	NCTUns	Computed optimal length of cluster in an intersection.	Cluster overheads and delays are increased.
RMAC [39]	Speed, location, and direction	Highway	Ns-2	Stable and less cluster reconfigurations	Collisions and unreliable.
VWCA [40]	Number of neighbors, the direction, the entropy, and the distrust value parameters	Not simulated	Not simulated	Predictability and reliability is increased.	Volatility of dynamic transmission range

Protocols	Cluster or CH	Scenario	Simulator	Pros	Cons
MOBIC [41]	Variance of relative mobility with each of its neighbour's.	Random	Ns-2	Reduces the cluster reconfiguration by delaying re-clustering for a certain period of time. To avoid accidental contacts between CHs.	Few neighbour nodes move differently, the method still results in dramatic increase in the variance.
AMACAD [42]	Relative distance, speed between neighbour's and distance between vehicle and destination.	Urban	Java JDeveloper 10G	Increases the cluster and CH lifetime.	Problem with knowing the final destination a priori as drivers usually do not use navigation system for known routes.
MCDRIVE [43]	First vehicle in the direction is elected as CH	Intersections	NCTUns	Cluster stability is improved in intersections.	Increased overhead and delay.
APROVE [44]	Minimum distance and minimum relative velocity between each CH and its CMs.	Highway	Ns-2, VanetMobiSim	Cluster overhead and re affiliation are reduced	It doesn't consider destination of vehicles. Not mention about CH election. Not suitable for intersections
ALM [45]	Variance in relative mobility	Box topology	SUMO, SIDE/ SMURPH	Considers relative mobility to increase stability.	No direction of movement and position is considered for cluster formation. Overhead increased.
DBC [46]	Connection graph density, link quality, traffic conditions, node reputation and movement prediction.	Urban	JiST/ SWANS++ VanetMobiSim	Suitable for both sparse and dense traffic.	The destination of vehicles, speed of vehicles is not taken into account that increases the overhead.
Maslekar [47]	Location and direction of vehicles.	Intersections	NCTUns	Cluster stability is improved in intersections	Overhead and delay increased.
Maslekar [48]	The direction which the vehicle will take after crossing intersection. The CH is at the front of cluster.	Intersections	NCTUns	Improved the influence of overtaking within the clusters. accurate density estimation within the clusters.	Overhead and delay increased.

Table 4. Comparison between various cluster based routing protocols.

http://dx.doi.org/10.5772/59041

3.4. CH election

CH selection is important to increase protocol reliability, scalability and delay. In some of CH selection algorithms proposed takes into account the destination of vehicles, including the current location, speed, direction, relative destination and final destination of vehicles as parameter to arrange the clusters. Many researchers have proposed CH election scheme based on ID. Each node is assigned a unique ID, and the node with the lowest ID in its two-hop neighborhood is elected to be the CH. Some algorithms calculate these ID based on the variance of relative mobility of a mobile node with each of its neighbors, where a small value of variance indicates the mobile node is moving relatively less than its neighborhood. Additionally, other approaches consider vehicles having a longer trip are more qualified for being elected as CHs.

A vehicle, which would travel longer time, is assigned higher priority; hence, at the very beginning of starting its travel, the expected travel time of a vehicle is calculated and announced using its desired driving speed and the geographic information system once its driver sets the destination. Te stability of the system is improved by electing the vehicles having a longer trip as the CHs. Furthermore, to avoid elected CHs losing connectivity with their neighbors very soon, the eligibility of a vehicle should decrease quickly when its velocity has big difference from the average speed. Thus, a vehicle with large speed deviation is assigned lower priority.

Another type of CH election scheme is based on connectivity level (estimating graph density), link quality (SNR), relative node position and the prediction of this position in the future, and node reputation. The vehicle which is near to that anchor point is elected as CH. Furthermore, some approaches assign generic weight to vehicles based on the position and other set of vehicle parameters like connectivity, mobility, RSS etc. The vehicle with the highest weight is elected as the CH amongst the neighbors. However, since the vehicles are highly dynamic in nature the position of the vehicles change very fast and hence may induce a computational overhead in calculating the weight associated with the vehicles.

In some of the clustering algorithm first vehicle entering into the cluster region is initialized as the CH. It changes from CH to CM due to the discovery of a closer CH, or until the last member of the cluster passes the intersection. However, CH stability is reduced due to distance between vehicle to intersection and due to different directions of vehicles. In some other schemes, the CH selection should resemble like a natural model of location references. CH re-election only occurs when two CHs move within range of one another for a certain contention interval. When a CM moves out of range of its CH, it joins any current CH in its neighborhood, or forms a new cluster. However, in the case in which few neighbor nodes move differently, the method still results in dramatic increase in the variance.

3.5. Cluster formation

Cluster formation is really important to avoid cluster reconfigurations. Some of the cluster formation techniques are based on position based clustering. In these types of protocols, each road is divided into cells and in each cell some anchor points are defined. The cluster structure is determined by the geographic position of the vehicle. Another type of position based algorithm is based on hierarchical and geographical data collection and dissemination mechanism. The cluster formation is based on the position of the vehicles at a particular segment instead of the individual positions. However, this type of protocols incurs more overheads for V2V and V2I communication. In some other approaches, each vehicle entering into the network collects the neighbor vehicles information, assuming precedence to each vehicle and polls each vehicle individually (according to precedence) to check whether it is CH or not and then joins the cluster. Also every vehicle in the network collects 2-hop neighbor's information along with 1-hop neighbor's information from the CH through periodic polling. These two information collection leads to more overhead in V2V communication.

Some clustering algorithms estimate the future mobility of nodes predicting the probability that the current neighborhood of a mobile node will remain the same. The drawback of the prediction method is the lack of accuracy in some cases. In some of clustering algorithms, the clusters are formed based on mobility metric and the signal power detected at the receiving vehicles on the same directed pathway. Through such method this type of protocol helps in forming stable clusters. However, it does not consider the losses prevalent in the wireless channel. In practical scenario effects of multi path fading are bound to affect the cluster formation method and thus the stability. These effects of multi path fading are taken into account in the density based clustering algorithm. The cluster formation is based on the weight metric which takes into consideration the link quality and the traffic conditions. It can be seen that the stability is improved compared to other approaches.

In some clustering approaches considers the behavior of the vehicles, using the speed and direction parameters.

The cluster formation is based on direction of vehicle at the approaching intersection. In other approach, cluster is formed based on distance and direction of vehicle it takes after crossing the junction. Some of the research enforces a weight cluster mechanism with a backup manager. These algorithms operate in similar way. Algorithms consider the position, direction, speed and range of the nodes to perform the algorithm. On the other hand, some takes into consideration the number of neighbors based on the dynamic transmission range, the direction of vehicles, the entropy, and the distrust value parameters. They works with an adaptive allocation of transmission range (AATR) technique, where hello messages and density of traffic around vehicles are used to adaptively adjust the transmission range among them. The destination of the vehicles is used as a parameter to arrange clusters.

In some approaches, the cluster formation interval is constant, which implies a synchronous creation of clusters. This does not allow for effective cluster reorganization. The directional based clustering algorithm are based on the following mobility metrics (a) moving direction (b) leadership duration (c) projected distance variation of all the neighboring vehicles over time. In practical scenario effects of multi path fading are bound to affect the cluster formation method and thus the stability. Some approaches take into account the destination of the vehicles to arrange the clusters and implements an efficient message mechanism to respond in real time and avoid global re-clustering. There might be a problem with knowing the final destination a priori as drivers usually do not use navigation system for known routes. Some algorithms are proposed for calculating the density of vehicles in a particular region around the junction. Moreover, other algorithms groups vehicles into clusters based on the competitive learning Hebb neural network. A suitable solution to prolong the cluster lifetime, stability, fairness, avoid congestion and overhead considering the vehicular behavior is essential.

3.6. Challenges of clustering

One of the numerous challenges clustering algorithms in VANETs is the mobile and dense communication topology. The main problem in clustering is the control overhead introduced to elect the CH and to maintain a stable cluster. The cluster structure assures the scalability of VANETs, where high mobility of the moving vehicles within the road causes lots of challenges to face. Location services might not provide the needed accuracy everywhere or will not be available at all so more work is needed on location independent clustering solutions. Providing highly accurate digital maps that are needed by some solutions presents a challenging task and could slow down the implementation so advantages and disadvantages of map based solutions should be researched.

In many papers the correlations between the transmission range and the VANETs density, packet transmit rate, packet size, data rate and channel conditions have been researched. However, the different network simulators should also be evaluated and presented with all the relevant parameters including MAC, transmission range, packet size, bit rate etc. Since each vehicle in the VANET has its own view of the network density and channel conditions, finding the optimal network parameters is difficult. The research should focus to the optimization of cluster size and transmission range that maintains a high VANET stability and reliability, increases the life time of a connection link, and at the same time decreases the time required for a safety message to reach its intended destination. Presented clustering protocols are designed for different aims e.g. overhead minimization, fast cluster creation, cluster stability, etc. The most important parameter among them is the cluster stability. Their tradeoffs and effects between them should also be analyzed and presented.

The vehicles with relatively high mobility, can pose difficulties for flat networks stability. Many of the presented protocols use metrics derived from the same input parameters where among them position and radio signal strength(RSS) are the most important. More research effort should be put in defining and ranking the aims that clusters and clustering protocols should try to achieve. One of the goals is to optimize the mean number of created clusters and the number of CHs at each time step.

For performance evaluations of clustering protocols common parameters used are cluster stability, CH election, cluster size, cluster delay, cluster reconfiguration and cluster overhead etc. These terms are quite generic so their definition and explanation with VANET specifics is needed to provide consistency between different researches. More focus should be put on evaluation of those common parameters to highlight the most useful ones, merge similar ones etc. This would help researchers to concentrate their research on extending and designing the most prospective ones. Fair comparison of different clustering protocols is a hard task due to non-existent standard testing procedures and scenarios so more work and standardization is needed in this area. The characteristics of different scenarios of VANETs and different parameters are explained in detail in later sections.

3.7. Scenarios

In highway scenarios, it is widely recognized that traffic generally follow a platoon pattern according to traffic flow theory. Vehicles in a platoon generally move with similar velocities and are likely to sustain a stable wireless communication in clusters. The clusters are independently controlled and dynamically reconfigured as the vehicles moving. Congestion can occur in highways during an accident so the clustered protocols should be designed to effectively reduce data congestion in high density scenarios, and satisfying QoS requirements. Furthermore, the design of cluster protocol should also consider the market penetration of vehicles enabled with OBU′s. In some cases, there can be a large number of vehicles in road that are not enabled with OBU′s. This creates a large gap between vehicles and resulting in poor communication. The future clustering protocols should consider all the characteristics of highway scenarios.

A large number of the available cluster based MAC and routing protocols are purposed for highway environments and does not address the various requirements of the city and urban traffic environments. In city environments, intersections play important roles for information exchange. The vehicle that crosses the intersection before actually receiving a message is defined as the unstable vehicle. As the intersection area is comparatively small and the probability of change of direction is very high, it will be risky to choose an unstable vehicle as the CH from these clusters. Moreover, during rush hours of day intersections are usually the bottlenecks.

Vehicles in intersection can take any of the direction Straight (S), Right (R), Left (L) and U-Turn (U) respectively. All the incoming vehicles of two road segments of intersection may be blocked by the red signal, whereas vehicles on the other two road segments flow until the green signal is on. When a vehicle crosses the intersection without having another vehicle arrive at the intersection, a disconnection may occur. Such a situation arises only when a fleet of vehicles has crossed the intersection and when another fleet of vehicles has not been arrived at the intersection. Based on the motion of vehicles, some approaches form clusters S, R, L and U on a particular lane. The created clusters consist of vehicles moving in the same direction. Within the same cluster the vehicles communicate with each other and elect a CH that is responsible for calculating the number of vehicles in its cluster. This information will help to avoid constant cluster reconfigurations and overhead by creating another cluster.

For intersection collision avoidance, the amount of traffic generated by vehicles can be determined by a number of factors such as the cluster size, the number of intersection per cluster, the number of vehicles per intersection per cluster, the size of messages, and the transmission interval.

3.8. Cluster size

The size of the cluster is a crucial parameter. To optimize the cluster size is very difficult especially in a highly mobile environment such as VANETs. One of the goals of optimal protocol is to optimize the number of CMs to decrease the end to end delay of messages. If the cluster size is decreased, the channel contention within each cluster decreases. However, the

http://dx.doi.org/10.5772/59041

number of CHs is increased, so that the resulting virtual network formed by these CHs will become more complex. There is then a tradeoff between the cluster size and the number of CHs. Cluster size is variable according to vehicle density, speed and required minimum bandwidth or QoS where parameters can be predefined or provided on the fly from vehicle sensors and application profiles.

The cluster size can be controlled by a predefined transmission range between a CH and its CMs. Optimal cluster size and hence the transmission range that maintains a high VANET reliability, stability and scalability, increases the life time of a communication link, and at the same time reduces the end to end delay for a safety message to reach its intended destination. Optimal cluster size is both related to the radio transmission power and vehicle traffic density. Therefore, cluster size may limit radio efficiency and throughput. For the cluster protocols, we have so far assumed that transmission power is fixed and is uniform across the VANET. There are different methods to reduce the cluster size by reducing transmission power. There is different power control protocols proposed but most of them are oscillating because of the fast varying vehicles densities in VANETs. Selection of optimal power control algorithm and vehicular densities will reduce the end to end delay, reliability and fairness.

Optimal cluster size is also determined by the correlation between spatial reuse of the medium (which leads to small numbers) and end to end delay minimization (which lead to large numbers). Other parameters also apply, such as geographical area and power consumption.

3.9. Stability

Stable clusters are important for a reliable and efficient information exchange. Stable clustering techniques decrease the control overhead of cluster reconfigurations and led to an efficient hierarchical VANET topology. The main condition for stability is the duration of residence's times of a cluster and its CHs. Stability is also defined by long CH lifetime, and long CM lifetime.

3.9.1. Cluster stability

Cluster stability is based on the selection of suitable CMs to ensure greater cluster lifetimes by reducing cluster re-configuration events. Cluster stability also depends on the different vehicle densities. To be able to form stable clusters of one hop vehicles, vehicular movements should be taken in to account. Speed and location data transfer is a usual procedure in most of the cluster-based routing protocols. Nevertheless, this needs two additional communication rounds (for speed, location and relative stability data transfer) and stationary assumption of vehicles prior to cluster creation. Cluster stability can be defined as the average number of cluster changes throughout the simulation and the percentage of time in which vehicles were CMs, represented as association time. In practical environment effects of multi path fading are bound to affect the cluster creation method and thus stability. In some cases, nodes in cluster are linked to cluster rather than CH. This increase furthers the cluster stability.

3.9.2. CH stability

The time during which a node is in the state of a CH determines stability. It is the mean time duration of the nodes, remaining its leadership role as CHs. Long CH lifetime implies, few changes and good stability. The information is disseminated by groups enhancing the communication delay, reliability, low data delivery and congestion issues, making the vehicular networks accurate and efficient. CH stability can be affected by different factors such as merging, distance between CHs, exit from the road etc. In VANETs, merging collisions can happen among vehicles moving in the same direction due to acceleration or deceleration, it is more likely to occur among vehicles moving in opposite directions (approaching each other) or between a vehicle and a stationary RSU since they approach each other with a much higher relative velocity as compared to vehicles moving in the same direction. The high mobility of the shifting nodes within the networks causes lots of challenges to face and affects stability.

If vehicles are changing their state very often in intersection scenarios and stay only for short times in the CH state, CH stability is low. In some of intersection based approaches the first vehicle to enter the intersection region in a particular direction is elected as CH to improve stability. Furthermore, some cluster-based routing algorithms, the selection of CH are based on willingness factor which defines the relative stability of a node. CH stability is also based on the threshold distance between the two CHs. Optimal distance between two CHs should be obtained.

3.9.3. Cluster delay

Cluster delay means the time required for sending one message from source to destination (it can be here from CM to the RSU or vice versa). The delay parameter is very crucial for safety applications. The end to end delay can be minimized by selecting proper cluster size, selection of proper MAC protocol to reduce the channel access time, selection of stable CH nearer to the RSU, a selection of proper routing algorithm between CH transmissions. The number of the formed clusters is important to reduce the end to end delay for message transmission.

3.9.4. Cluster reconfiguration

The frequent cluster reconfiguration generates tremendous communication load, which significantly reduces available bandwidth for message dissemination. Cluster reconfiguration is needed in some cases when the CH leaves the group or numbers of CMs are below the threshold or the distance between two CH is below the threshold. In some approaches, if the distance between two CH nodes is detected less than the particular threshold, the cluster with fewer CMs is dismissed to reduce communication overheads while it's CMs join other clusters. One can expect that a larger dismiss threshold leads to a higher rate of CH changes and higher probability of cluster reconfiguration. The threshold determines the rate of cluster reconfiguration, and also, depends on the radio transmission range and vehicular densities. Larger transmission provides longer distance for CHs to detect each other, and therefore, more frequent cluster reconfigurations occur. Additionally, some algorithm elects backup CH to avoid cluster reconfigurations. However, most of the protocols are not fit for different traffic

http://dx.doi.org/10.5772/59041

situations. The aim should be to design protocol with less cluster reconfiguration in various scenarios.

3.9.5. Cluster throughput

Data rate transfer that gives the total number of received packets at the destination out of total transmitted packets. An access collision happens when two or more CHs within two hops of each other attempt to acquire the same available time slot.

3.9.6. Cluster overhead

Clustering requires explicit clustering-related information exchanged between node pairs. Clusters cannot be formed or maintained by non-clustering-related messages, such as routing information or data packets. The main challenge in clustering is the communication overhead introduced to formation and maintenance of a stable cluster, and elects its stable CH. Most of the recently proposed protocols discuss mainly on how CHs are selected. The control overhead for the creation and reconfigurations of clusters have not been considered completely. There have been not many papers that analyze analytically the control overhead incurred in hierarchical routing. Furthermore, the overhead is bound by a constant per vehicle per time step, avoiding expensive re-clustering chain reactions; hence, this overhead increases with the number of nodes. Since a CH acts as a coordinator in a cluster, if it is absent for any reason, the clustering architecture has to be reconfigured; this will significantly increase the message overhead.

Communication complexity represents the total amount of clustering-related message exchanged for the cluster formation. For clustering schemes with ripple effect, the communication complexity for the re-clustering in the cluster maintenance phase may be the same as that in the cluster formation phase. But for those with no ripple effect, the communication complexity of re-clustering should be much lower. From analysis of different clustering protocols, we believe that a more efficient way to form a stable network structure, with reduced overhead, are that a vehicle should be associated to a cluster and not to a CH. Indeed, replacing CH is considered only as an incremental update and does not require a whole reconfiguration of the cluster structure; this will definitively increase the lifetime of the clustering architecture. The resulting clusters are stable and exhibit long average CM duration, long average CH duration, and low average rate of CH changes. The cluster creation and maintenance overhead should be calculated to be compared with non-clustering algorithms in terms of the reliability, fairness, and scalability of the algorithms. By optimizing cluster stability, cluster reconfiguration, number of clusters and cluster size can reduce the overhead caused in clustering.

4. Conclusions

In this chapter, we have surveyed in-depth of the challenges of reliable communication for cooperative ad hoc networks especially VANETS. First we have provided state of the art of ad hoc networks and various types of ad hoc networks.

In a scenario where nodes are moving fast and the topology of the network is changing continuously, the big challenge is to keep connected all the nodes and give all of the them resources to transmit and receive information in real time. In VANETs, dissemination algorithms provide to the drivers mechanisms to be aware in real-time of events that are happening in their surroundings: traffic and road conditions, closure and detour information, incident information, emergency alerts, and driver advisories. Clustering is an approach that divides the nodes of the network in groups of vehicles according to common characteristics as their position or speed, in order to create a more robust and scalable network. This structure can be a realistic approach to support reliable and scalable multihop communications in a mobile network as a VANET

This chapter focused on identifying the research trend of the cluster based MAC and routing techniques that have been recently proposed for V2I and V2V communications. Furthermore, we discussed the advantages and disadvantages of various MAC protocols that have been developed recently. Moreover, we have presented a comprehensive review of the cluster based routing protocols for inter-vehicle and vehicle-to-infrastructure communication. Cluster based routing protocol is the most appropriate technique for developing reliable, scalable and predictable routing protocols in VANETs. However, due to the distinctive attributes of V2V and V2I communications, it raises several open issues and areas for research, such as fairer usage of network resources and channel access. Because of varying vehicle density and varying speed of vehicles makes communication reliability, a challenging issue.

Our research group is focusing on developing data dissemination and cluster based protocols by identifying common characteristics and parameters to improve cluster lifetime, communication link among vehicles, and channel access. We have also discussed some important issues that must be addressed for safety and non-safety applications. Future protocols need to effectively consider these problems while fully exploiting the distinctive distributive and ad hoc nature of these networks to meet real time applications.

Acknowledgements

The authors would like to thank the EU Intelligent Cooperative Sensing for Improved traffic efficiency (ICSI) project (FP7-ICT-2011-8) for its support in the development of this work.

Author details

Unai Hernandez-Jayo*, Aboobeker Sidhik Koyamparambil Mammu and Idoia De-la-Iglesia

*Address all correspondence to: unai.hernandez@deusto.es

Deusto Institute of Technology, DeustoTech Mobility, Bilbao Bizkaia, Spain

http://dx.doi.org/10.5772/59041

References

[1] Verma, L.; Lee, S.S., Proliferation of Wi-Fi: Opportunities in CE ecosystem. Consumer Communications and Networking Conference (CCNC), 2011 IEEE, vol., no., pp. 213,217, 9-12 Jan. 2011

[2] Saleh Ali K. Al-Omari and Putra Sumari. An Overview of Mobile Ad Hoc Networks for the Existing Protocols and Applications. Journal on Applications of Graph Theory in Wireless Ad hoc Networks and Sensor Networks, Vol.2, No.1, March 2010

[3] Corson, S. and Macker, J. Mobile Ad Hoc Networking (MANET): Routing Protocol Performance Issues and Evaluation Considerations. RFC Editor, 1999.

[4] Kevin Ashton. The Internet of Things and other things. http://kevinjashton.com/ (accessed 24 June 2014).

[5] IoT Interview Series: 5 questions with Inaki Vazquez. http://postscapes.com/ (accessed 24 June 2014).

[6] Nithya Darisini, P.S.; Kumari, N.S. A survey of routing protocols for VANET in urban scenarios. Pattern Recognition, Informatics and Mobile Engineering (PRIME), 2013 International Conference on, vol., no., pp.464,467, 21-22 Feb. 2013

[7] Michoud, R.; Orozco, A.M.; Llano, G. Mobile ad-hoc routing protocols survey for the design of VANET applications. Intelligent Transportation Systems Symposium (CITSS), 2012 IEEE Colombian, vol., no., pp.1,6, 30-30 Aug. 2012

[8] Sherali Zeadally, Ray Hunt, Yuh-Shyan Chen, Angela Irwin, Aamir Hassan. Vehicular ad hoc networks (VANETS): status, results, and challenges. Telecommunication Systems, Springer, Vol.50, Issue 4, pp 217-241, Aug. 2012

[9] ETSI TR 102 638. Intelligent Transport Systems (ITS); Vehicular Communications; Basic Set of Applications; Definitions. 2009-06

[10] Verma, L.; Lee, S.S. Proliferation of Wi-Fi: Opportunities in CE ecosystem. Consumer Communications and Networking Conference (CCNC), 2011 IEEE, vol., no., pp. 213,217, 9-12 Jan. 2011

[11] IEEE Guide for Wireless Access in Vehicular Environments (WAVE)-Architecture," IEEE Std 1609.0-2013, vol., no., pp.1,78, March 5 2014

[12] Yang, X., Liu, J., Vaidya, N. F., & Zhao, F. (2004, August). A vehicle-to-vehicle communication protocol for cooperative collision warning. In Mobile and Ubiquitous Systems: Networking and Services, 2004. MOBIQUITOUS 2004. The First Annual International Conference on (pp. 114-123). IEEE.

[13] Ferrari, G., Busanelli, S., Iotti, N., & Kaplan, Y. (2011, August). Cross-network information dissemination in VANETs. In ITS Telecommunications (ITST), 2011 11th International Conference on (pp. 351-356). IEEE.

[14] Y. Cao, S. Xie: A Position-based Beaconless Routing Algorithm for Mobile Ad Hoc Networks, in Proceedings of International Conference on Communications, Circuits and Systems, Vol. 1, IEEE, 2005, pp. 303-307

[15] Du, D.Z: Ad hoc wireless Networking, Kluwer, 2003, pp. 103-136 http://www.site.uottawa.ca/~ivan/routing-survey.pdf

[16] M. Mauve, J. Widmer, H. Hartenstein: A Survey on Position-based Routing in Mobile Ad-Hoc Networks, IEEE Network, Vol. 15, No. 6, 2001, pp. 30-39

[17] V. Giruka, M. Singhal: Angular Routing Protocol for Mobile Ad-hoc Networks, in Proceedings of 25th IEEE International Conference on Distributed Computing Systems Workshops (ICDCSW'05), 2005, pp. 551-557

[18] S. Giordano, I. Stojmenovic, L. Blazevic: Position-based Routing Algorithms for Ad Hoc Networks: A Taxonomy. In Cheng, X., Huang, X., Du, D.Z: Ad hoc wireless Networking, Kluwer, 2003, pp. 103-136 http://www.site.uottawa.ca/~ivan/routing-survey.pdf

[19] C. Maihöfer. A survey of geocast routing protocols. IEEE Communications Surveys & Tutorials, vol. 6, no. 2, pp. 32–42, 2004.

[20] Little. T.D.C., Agarwal. A. An information propagation scheme for VANETs. IEEE, nº 13-15 Sept. 2005, pp. 155-160, 2005.

[21] R. Ramanathan, M. Steenstrup. Hierarchically-organized multihop mobile wireless networks for quality-of-service support. Mob. Netw.Appl, vol. 3, nº 1, pp. 101-119, 1998.

[22] C. Wai y C. Shengwei. Ad hoc peer-to-peer network architecture for vehicle safety communications. Communications Magazine, IEEE, vol. 43, nº April, pp. 100-107, 2005.

[23] Chandra Rathore. N, Verma. S, Tomar. G.S. CMAC: A cluster based MAC protocol for VANETs. Computer Information Systems and Industrial Management Applications (CISIM), 8-10 Oct. 2010.

[24] R. Tomar, S. Verma, G. Tomar. Cluster based RSU centric channel access for VANETs. Lecture Notes in Computer Science, Springer Berlin Heidelberg, vol. 7410, nº XVII, p. 150–171, 2013.

[25] E. Dror, C. Avin, Z. Lotker. Fast randomized algorithm for hierarchical clustering in vehicular ad-hoc networks. Ad Hoc Networking Workshop (Med-Hoc-Net), 2011 The 10th IFIP Annual Mediterranean, 2011.

[26] Z. Rawashdeh, S. Mahmud. Media access technique for cluster-based vehicular ad hoc networks. Vehicular Technology Conference, 2008, VTC 2008-Fall. IEEE 68th, Sept., pp. 1–5., 2008.

[27] X. Zhang, H. Su, H.H. Chen. Cluster-based multi-channel communications protocols in vehicle ad hoc networks. Wireless Communications IEEE, p. 13 (5) (October) 44–51., 2006.

[28] Y. Gunter, B. Wiegel, H. Grossmann. Cluster-based medium access scheme for vanets. Intelligent Transportation Systems Conference. ITSC 2007. IEEE, 30 2007-Oct. 3, pp. 343–348., 2007.

[29] Y. C. Lai, P. Lin, W. Liao, C.M. Chen. A region-based clustering mechanism for channel access in vehicular ad hoc networks. Selected Areas Communications, IEEE Journal, p. on 29 (1) (January) 83–93., 2011.

[30] M. Almalag, S. Olariu, M. Weigle. TDMA cluster-based MAC for VANETs (TC_MAC). World of Wireless, Mobile and Multimedia Networks (WoWMoM), IEEE International Symposium, June, pp. 1–6., 2012.

[31] O. Kayis, T. Acarman. Clustering formation for inter-vehicle communication. Intelligent Transportation Systems Conference ITSC, 2007.

[32] F. Borgonovo, A. Capone,M. Cesana, L. Fratta. Adhoc MAC: new MAC architecture for ad hoc networks providing efficient and reliable point-to-point and broadcast. Wirel. Netw, p. 10 (4) (2004) 359–366., 2004.

[33] S. Bana, P. Varaiya. Space division multiple access (SDMA) for robust ad hoc vehicle communication networks. Intelligent Transportation Systems Proceedings IEEE, p. 962–967, 2001.

[34] H. Omar, W. Zhuang, L. Li. VEMAC: A novel multichannel MAC protocol for vehicular ad hoc networks. Computer Communications Workshops (INFOCOM WKSHPS), 2011 IEEE Conference on, April, pp. 413–418., 2011.

[35] N. Lu, Y. Ji, F. Liu, X. Wang. A dedicated multi-channel MAC protocol design for VANET with adaptive broadcasting. Wireless Communications and Networking Conference (WCNC), 2010 IEEE, 2010, pp. 1–6, 2010.

[36] K. Bilstrup, E. Uhlemann y E. Strom. Scalability Issues of the MAC Methods STDMA and CSMA of IEEE 802.11p When Used in VANETs. de Communications Workshops (ICC), 2010 IEEE International Conference on, vol., no., pp.1,5, 23-27 May, 2010.

[37] F. Yu, S. Biswas. Self-configuring tdma protocols for enhancing vehicle safety with DSRC based vehicle-to-vehicle communications. Selected Areas in Communications, IEEE Journal, vol. on 25 (8) (2007) 1526–1537, 2007.

[38] M. Venkata, M. Pai, R. Pai, J. Mouzna. Traffic monitoring and routing in VANETs-a cluster based approach. ITS Telecommunications (ITST), 11th International Conference on, pp. 27–32., 2011.

[39] R. T. Goonewardene, F. Ali, E. Stipidis. Robust mobility adaptive clustering scheme with support for geographic routing for vehicular ad hoc. Intelligent Transport Systems, IET 3 (2) (June) 148–158., 2009.

[40] A. Daeinab, A. G.P. Rahbar, A. Khademzadeh. Vwca: An efficient clustering algorithm in vehicular ad hoc networks. Journal of Network and Computer Applications 34 (1)207 – 222., 2011.

[41] P. Basu, N. Khan, T. D. C. Little. A mobility based metric for clustering in mobile ad hoc networks. Distributed Computing Systems Workshop, International Conference on, pp. 413–418., 2001.

[42] Morales. M.M.C, Choong-seon. Hong, Young. Cheol Bang. An Adaptable Mobility-Aware Clustering Algorithm in vehicular networks. Network Operations and Management Symposium (APNOMS), 2011 13th Asia-Pacific, 2011.

[43] N. Maslekar, J. Mouzna, H. Labiod, M. Devisetty, M. Pai. Modified c-drive: Clustering based on direction in vehicular environment. Intelligent Vehicles Symposium (IV), 2011 IEEE, June, pp. 845–850, 2011.

[44] C. Shea, B. Hassanabadi, S. Valaee. Mobility-based clustering in VANETs using affinity propagation. Global Telecommunications Conference, GLOBECOM 2009. IEEE, 30 2009-Dec. 4, pp. 1–6, 2009.

[45] E. Souza, I. Nikolaidis, P. Gburzynski. A new aggregate local mobility clustering algorithm for VANETs. Communications (ICC) IEEE International Conference on, 2010, pp. 1–5, 2010.

[46] S. Kuklinski, G. Wolny. Density based clustering algorithm for VANETs. de Testbeds and Research Infrastructures for the Development of Networks Communities and WorkshopsTridentCom 2009. 5th International Conference on, April, pp. 1–6, 2009.

[47] N. Maslekar, M. Boussedjra, J. Mouzna, L. Houda. Direction based clustering algorithm for data dissemination in vehicular networks. Vehicular Networking Conference (VNC), 2009 IEEE, Oct., pp. 1–6, 2009.

[48] N.Maslekar,M. Boussedjra, J.Mouzna, H. Labiod. A stable clustering algorithm for efficiency applications in VANETs. de Wireless Communications and Mobile Computing Conference (IWCMC), 2011 7th International, pp. 1188–1193, 2011.